VECTOR ANALYSIS FOR MATHEMATICIANS SCIENTISTS AND ENGINEERS

S. SIMONS

Lecturer in Mathematics
Queen Mary College
University of London

PERGAMON PRESS
OXFORD · LONDON · EDINBURGH · PARIS · FRANKFURT

THE MACMILLAN COMPANY
NEW YORK

PERGAMON PRESS LTD.   Headington Hill Hall, Oxford
                        4 & 5 Fitzroy Square, London W.1

PERGAMON PRESS   2 & 3 Teviot Place, Edinburgh 1
(SCOTLAND) LTD.

THE MACMILLAN COMPANY   60 Fifth Avenue,
                        New York 11, N.Y.

COLLIER-MACMILLAN   132 Water Street South,
CANADA LTD.   Galt, Ontario, Canada

GAUTHIER-VILLARS ED.   55 Quai des Grands-Augustins, Paris 6

PERGAMON PRESS G.m.b.H. Kaiserstrasse 75, Frankfurt am Main

FEDERAL PUBLICATIONS LTD. Times House,
                        River Valley Road, Singapore

SAMCAX BOOK SERVICES LTD. Queensway, P.O. Box 2720,
                        Nairobi, Kenya

                        Printed in Great Britain by
Set in 10 on 12 pt. Times   BELL & BAIN LTD., GLASGOW

Contents

v

Preface

THE METHODS of vector algebra and calculus are now accepted as an essential part of the mathematical equipment of Physicists, Engineers and others working in the physical sciences. It is the aim of this book to supply the necessary understanding of these methods to the extent that the student will readily follow those works which make use of them, and further, will be able to employ them himself in his own branch of science. To this end, the new concepts and methods introduced are illustrated by examples drawn from fields with which the student is familiar, and a large number of both worked and unworked exercises are provided. Detailed discussions of the finer mathematical points arising in the course of the work are, however, omitted.

Notable features of the treatment include a discussion of rotational invariance of the grad, div and curl operators, and a chapter on the direct evaluation of line, surface and volume integrals. The power of vector methods is finally illustrated by a treatment of electricity, in which the main thread of electrical theory is traced, from the basic experimental data up to Maxwell's equations.

1

Introduction to Vectors

1.1 WHAT IS A VECTOR?

An examination of physical quantities shows that they may be divided into two classes depending on what is required to specify them completely. Thus mass, volume, temperature, electric charge, etc., are all completely specified (once the unit is given) by means of a *single* number, e.g. a mass of *five* pounds, a temperature of *twenty-three* degrees Centigrade, etc. On the other hand, a statement such as " the velocity of a car is forty-six miles per hour " is certainly incomplete in the details it gives us of the car's motion. In order to know this completely we should have to be told the *direction* of movement—" the velocity of a car is *forty-six* miles per hour in a direction *North* 40° *East* ". Other physical quantities such as force, momentum, electric field, etc., are similar in that their complete specification requires a statement of direction in addition to one of magnitude. Such quantities are called *vectors* in contradistinction to those mentioned earlier and *not* involving direction, which are called *scalars*.

> A VECTOR is a quantity having both *magnitude* and *direction*.
>
> A SCALAR is a quantity having *magnitude* but no direction.

1

The importance of a study of vectors lies in the fact that since they correspond to certain physical quantities, calculations involving these quantities can be considerably simplified and results more easily apprehended by the use of vector analysis than by other methods; this will be amply illustrated in later chapters.

1.2   REPRESENTATION OF VECTORS

Since a vector has both magnitude and direction, it may be completely represented by a straight line *OP* in the direction of

Fig. 1.

the vector and of length corresponding to the magnitude of the vector according to a convenient scale. The *sense* of the vector, i.e. from *O* to *P*, is shown by means of an arrow as in Fig. 1. Such a vector would be referred to as $\overrightarrow{OP}$, or by a single letter in bold type, e.g. **F**. The scalar magnitude of the vector is denoted by $|\overrightarrow{OP}|$ or $|\mathbf{F}|$, while if no ambiguity is possible, the simpler notation *OP* or *F* may be used. A vector parallel to **F** and of magnitude k times the magnitude of **F** is denoted by $k\mathbf{F}$, while $-\mathbf{F}$ denotes a vector equal in magnitude to **F** but in the opposite direction; in the notation of Fig. 1,

$-\mathbf{F} = \overrightarrow{PO}$. A vector parallel to $\mathbf{F}$ and of unit magnitude is called a unit vector parallel to $\mathbf{F}$, and is represented by $\hat{\mathbf{F}}$; it is clear that with these conventions, $\mathbf{F} = F\hat{\mathbf{F}}$. It should be emphasised here that a vector is unaltered by pure displacement, and thus the two vectors $\overrightarrow{AB}$ and $\overrightarrow{CD}$ of Fig. 2, being parallel and of equal magnitude, are equal vectors $\mathbf{G}$.

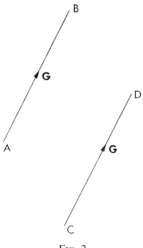

FIG. 2.

1.3 ADDITION AND SUBTRACTION OF VECTORS

We are quite used to adding scalar quantities: for example, a mass of 7 lb plus a mass of 4 lb equals a mass of 11 lb. This direct addition of magnitudes cannot, however, be applied directly to vectors owing to the directional element involved. In fact, our common ideas on scalar addition do not lead logically to any definition of vector addition, and we are therefore at liberty to define this in any convenient way. Now it is a familiar result of elementary mechanics that the resultant of any two vectorial physical quantities such as velocity or force is obtained as the third side of a triangle whose other two sides represent the given velocities or forces.

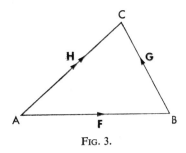

Fig. 3.

We therefore use this construction to *define* the sum of the vectors $\overrightarrow{AB}$ and $\overrightarrow{BC}$ shown in Fig. 3 as follows:

> The SUM of the two vectors $\overrightarrow{AB}$ and $\overrightarrow{BC}$ is the vector $\overrightarrow{AC}$.

This result may be written

$$\overrightarrow{AB} + \overrightarrow{BC} = \overrightarrow{AC}$$

or with the single letter labelling of Fig. 3,

$$\mathbf{F} + \mathbf{G} = \mathbf{H}.$$

It is clear that with this definition, the resultant of two velocities or forces is their sum. It should also be noticed that if the vectors $\mathbf{F}$ and $\mathbf{G}$ are parallel, then this definition is equivalent to scalar addition of the magnitudes of the vectors.

Having defined the vector $-\mathbf{F}$ in the previous section, the difference between two vectors **F** and **G**, written **F** − **G**, is defined by

$$\mathbf{F} - \mathbf{G} = \mathbf{F} + (-\mathbf{G});\qquad(1)$$

i.e. the difference between **F** and **G** is the sum of **F** and −**G**. This is illustrated in Fig. 4.

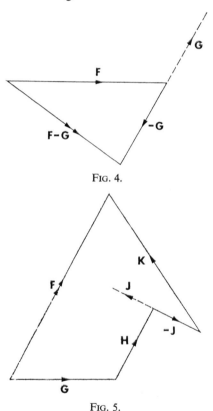

Fig. 4.

Fig. 5.

It is clear that repeated applications of the above definitions may be used to obtain sums and differences of several vectors. This is illustrated in Fig. 5 where the vector **F** given by

$\mathbf{F} = \mathbf{G} + \mathbf{H} - \mathbf{J} + \mathbf{K}$ is obtained. It is left as an exercise to the student to show that when adding several vectors, the order and grouping of the vectors is immaterial, e.g. $\mathbf{a} + \mathbf{b} = \mathbf{b} + \mathbf{a}$, and $(\mathbf{a} + \mathbf{b}) + \mathbf{c} = \mathbf{a} + (\mathbf{b} + \mathbf{c})$. These results show that as far as addition and subtraction is concerned, vectors may be manipulated in the same way as ordinary algebraic scalars.

Exercise

Given the three vectors **G, H, K**, shown in Fig. 5, construct
  (i) $\mathbf{G} - (\mathbf{H} + \mathbf{K})$,
  (ii) $2\mathbf{G} + \frac{1}{2}(\mathbf{H} - \mathbf{K})$.

1.4 SIMPLE GEOMETRICAL APPLICATIONS

In order to illustrate the concepts introduced in the previous sections, we now consider some simple applications to geometry, and for this purpose we define the *position vector* **r** of a general point P relative to a fixed origin O as the vector

$$\mathbf{r} = \overrightarrow{OP}.$$

Suppose now that we know that $a\mathbf{r} + b\mathbf{s} = 0$ for two scalars a and b, and two position vectors **r** and **s**. What can be deduced from this relation? It follows immediately that $\mathbf{r} = -(b/a)\mathbf{s}$ if $a \neq 0$, and therefore **r** and **s** must be parallel, since **r** is a scalar multiple of **s**. **r** and **s** are not necessarily parallel only if $a = 0$, and when this is so the original equation shows that $b = 0$. We thus conclude that if $a\mathbf{r} + b\mathbf{s} = 0$, *either* **r** is parallel to **s** *or* $a = b = 0$. This result may be used to show that the diagonals of a parallelogram bisect each other, for let $ABCD$ (Fig. 6) represent a parallelogram whose sides are the vectors **r** and **s**, and whose diagonals intersect at P. Then clearly $\overrightarrow{BD} = \mathbf{s} - \mathbf{r}$ since $\mathbf{r} + \overrightarrow{BD} = \mathbf{s}$, and $\overrightarrow{AC} = \mathbf{r} + \mathbf{s}$. Thus $\overrightarrow{BP} = a(\mathbf{s} - \mathbf{r})$ for some scalar a and

$\overrightarrow{AP} = b(\mathbf{r} + \mathbf{s})$ for some scalar b. Now since $\overrightarrow{AB} + \overrightarrow{BP} = \overrightarrow{AP}$, it follows that $\mathbf{r} + a(\mathbf{s} - \mathbf{r}) = b(\mathbf{r} + \mathbf{s})$, whence

$$\mathbf{r}(1 - a - b) + \mathbf{s}(a - b) = 0.$$

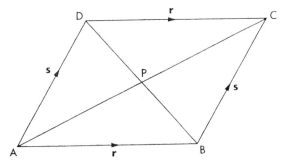

Fig. 6.

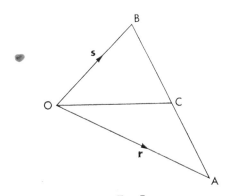

Fig. 7.

Since $\mathbf{r}$ and $\mathbf{s}$ are not parallel, the result established earlier in this section yields the result $1 - a - b = 0$ and $a - b = 0$, which have the solution $a = b = \frac{1}{2}$. Hence the diagonals bisect each other.

Again, let **r**, **s** be the position vectors $\overrightarrow{OA}$ and $\overrightarrow{OB}$ shown in Fig. 7. If C is the mid-point of AB, what will $\overrightarrow{OC}$ be in terms of **r** and **s**? Clearly

$$\overrightarrow{CO} + \mathbf{r} = \overrightarrow{CA} = -\overrightarrow{CB} = -(\overrightarrow{CO} + \mathbf{s})$$

whence $2\overrightarrow{CO} = -(\mathbf{r} + \mathbf{s})$, and therefore

$$\overrightarrow{OC} = \tfrac{1}{2}(\mathbf{r} + \mathbf{s}). \tag{2}$$

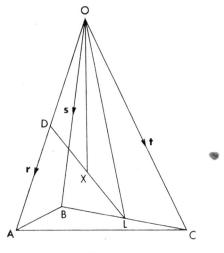

FIG. 8.

This result may be used to show that the lines joining the midpoints of opposite sides of a tetrahedron are concurrent, for let **r**, **s** and **t** be the position vectors $\overrightarrow{OA}$, $\overrightarrow{OB}$, $\overrightarrow{OC}$ corresponding to the edges of the tetrahedron $OABC$ shown in Fig. 8. Then, if D and L are the mid-points of the edges OA and BC respectively, $\overrightarrow{OD} = \tfrac{1}{2}\mathbf{r}$ and $\overrightarrow{OL} = \tfrac{1}{2}(\mathbf{s} + \mathbf{t})$ from the result (2). If X is the mid-point of LD, further application of eqn. (2) gives

$$\overrightarrow{OX} = \tfrac{1}{2}[\tfrac{1}{2}\mathbf{r} + \tfrac{1}{2}(\mathbf{s} + \mathbf{t})] = \tfrac{1}{4}(\mathbf{r} + \mathbf{s} + \mathbf{t}).$$

Since this expression for $\overrightarrow{OX}$ is symmetric in $\mathbf{r}$, $\mathbf{s}$ and $\mathbf{t}$, the same result will be obtained for the mid-point of the mid-points of any other pair of edges; i.e. the lines joining the mid-points of opposite edges are all concurrent at X. It should be noticed that the above vector treatment is not

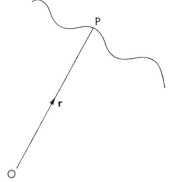

FIG. 9 .

more complicated because it is in three dimensions. As will be seen in more detail later, an important advantage of vector analysis in general is that it is equally adapted for work in two or three dimensions, and the latter does not introduce additional complications.

Vector Equation of a Curve or Surface

It is possible to give the " vector equation " of a curve or surface. By this we mean that the position vector $\overrightarrow{OP}\,(=\mathbf{r})$ is taken from a fixed origin O to a point P on the curve or surface as shown in Fig. 9, and the relation satisfied by **r** for all points on the curve or surface is called its vector equation. Consider, for example, a straight line passing through points A and B such that $\overrightarrow{OA} = \mathbf{s}$ and $\overrightarrow{OB} = \mathbf{t}$ as shown in Fig. 10.

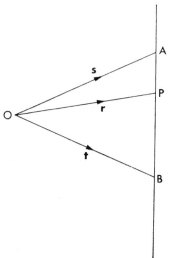

FIG. 10.

Then for any point P on the line

$$\overrightarrow{OP} = \mathbf{r} = \mathbf{s} + \overrightarrow{AP} = \mathbf{s} + b\overrightarrow{AB},$$

where $b = AP/AB$. Hence, since $\overrightarrow{AB} = \mathbf{t} - \mathbf{s}$,

$$\mathbf{r} = \mathbf{s} + b(\mathbf{t} - \mathbf{s}) = (1 - b)\mathbf{s} + b\mathbf{t}. \tag{3}$$

Equation (3) is the required vector equation of the line involving the parameter b, since for a suitable value of b, eqn. (3) is satisfied for all points on the line.

Exercise

Use eqn. (3) to show that if $ABCD$ is a parallelogram and E is the mid-point of AB, then DE and AC trisect each other.

[*Hint*. Obtain the equation of DE and AC, and hence find their point of intersection.]

1.5 COMPONENTS OF A VECTOR

Consider a given vector $\mathbf{F}$ $(= \overrightarrow{OR})$, together with another two non-parallel coplanar vectors $\mathbf{u}$ $(= \overrightarrow{OU})$ and $\mathbf{v}$ $(= \overrightarrow{OV})$ as shown in Fig. 11. We shall now see that it is always possible to choose scalars a and b so that

$$\mathbf{F} = a\mathbf{u} + b\mathbf{v}. \tag{4}$$

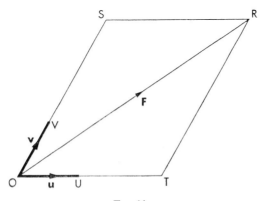

FIG. 11.

To prove this, draw lines RS, RT parallel to OU, OV respectively to meet OV, OU (produced if necessary) at S and T respectively. Then, since $OSRT$ is a parallelogram,

$$\overrightarrow{OR} = \overrightarrow{OS} + \overrightarrow{SR} = \overrightarrow{OS} + \overrightarrow{OT}. \tag{5}$$

Also, since O, S and V are collinear, $\overrightarrow{OS} = b\overrightarrow{OV}$ for some scalar b, and similarly $\overrightarrow{OT} = a\overrightarrow{OU}$ for some scalar a. Hence, substituting into eqn. (5),

$$\overrightarrow{OR} = a\overrightarrow{OU} + b\overrightarrow{OV}, \tag{6}$$

which is identical with eqn. (4). It is clear from the above construction that the values of a and b are unique.

Now, suppose that **F** and **p**, **q**, **s** are four non-coplanar vectors. Then, by construction of a parallelepiped with body diagonal **F**, and edges parallel to **p**, **q** and **s** as in Fig. 12, it may readily be shown by an argument analogous to that given above, that it is possible to find unique scalars a, b, c so that

$$\mathbf{F} = a\mathbf{p} + b\mathbf{q} + c\mathbf{s}. \tag{7}$$

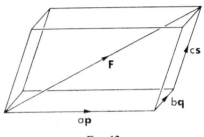

Fig. 12.

F is then said to be *resolved* into *components* a**p**, b**q**, c**s** parallel to **p**, **q**, **s** respectively.

Cartesian Components

For many purposes, it is convenient to take components along three mutually perpendicular directions OX, OY, OZ where these are the usual three-dimensional cartesian axes as shown in Fig. 13. If we select unit vectors **i** along OX, **j** along OY, and **k** along OZ, then we can find points A, B, C along OX, OY, OZ respectively, so that

$$\mathbf{F} = OA\mathbf{i} + OB\mathbf{j} + OC\mathbf{k}. \tag{8}$$

Putting $OA = F_x$, $OB = F_y$, and $OC = F_z$, since these

symbols show the vector; (i.e. **F**) and the axis (x, y or z) to which they refer, we have

$$\mathbf{F} = F_x\mathbf{i} + F_y\mathbf{j} + F_z\mathbf{k}. \tag{9}$$

The scalar quantities F_x, F_y, F_z are called the *resolutes* of **F** along the x, y, z axes respectively, and it is clear that

$$F_x = F \cos \theta_x, \ F_y = F \cos \theta_y, \ F_z = F \cos \theta_z, \tag{10}$$

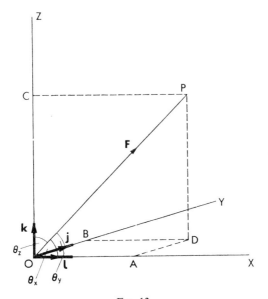

Fig. 13.

where θ_x, θ_y, θ_z are the angles made by **F** with OX, OY, OZ respectively. Applying Pythagoras' theorem to Fig. 13 readily yields

$$OP^2 = OA^2 + AD^2 + DP^2 = OA^2 + OB^2 + OC^2;$$

i.e.

$$F^2 = F_x{}^2 + F_y{}^2 + F_z{}^2. \qquad (11)$$

The angles θ_x, θ_y, θ_z appearing in eqn. (10) are not independent, since substituting from eqn. (10) into eqn. (11) yields

$$F^2 = F^2 \cos^2 \theta_x + F^2 \cos^2 \theta_y + F^2 \cos^2 \theta_z,$$

i.e.

$$\cos^2 \theta_x + \cos^2 \theta_y + \cos^2 \theta_z = 1. \qquad (12)$$

It should be emphasised that the resolute of $\mathbf{F}$ in a given direction, e.g. F_x is a scalar, while its component in the same direction, i.e. $F_x\mathbf{i}$ is a vector.

It is clear that for any point P in space with position vector $\mathbf{r}$ ($= \overrightarrow{OP}$), we have

$$r_x = x, \quad r_y = y, \quad r_z = z, \qquad (13)$$

where x, y, z are the cartesian coordinates of P.

Given two vectors $\mathbf{F}$ ($= F_x\mathbf{i} + F_y\mathbf{j} + F_z\mathbf{k}$) and $\mathbf{G}$ ($= G_x\mathbf{i} + G_y\mathbf{j} + G_z\mathbf{k}$), their sum will be

$$\mathbf{F} + \mathbf{G} = F_x\mathbf{i} + F_y\mathbf{j} + F_z\mathbf{k} + G_x\mathbf{i} + G_y\mathbf{j} + G_z\mathbf{k}$$
$$= (F_x + G_x)\mathbf{i} + (F_y + G_y)\mathbf{j} + (F_z + G_z)\mathbf{k}; \qquad (14)$$

i.e. the addition of two or more vectors corresponds to adding their resolutes in each of the three directions.

Suppose that $\mathbf{F} = \mathbf{G}$. This implies $\mathbf{F} - \mathbf{G} = 0$, and therefore

$$(F_x - G_x)\mathbf{i} + (F_y - G_y)\mathbf{j} + (F_z - G_z)\mathbf{k} = 0. \qquad (15)$$

Now, it is clear that for a vector to be zero, each of its three resolutes must be independently zero, and so eqn. (15) implies that

$$F_x = G_x, \quad F_y = G_y, \quad F_z = G_z;$$

i.e. if two vectors are equal, their resolutes in each direction are equal.

Worked examples

(1) Obtain the unit vector parallel to the sum of

$$\mathbf{F} = 3\mathbf{i} - 5\mathbf{j} + 8\mathbf{k} \quad \text{and} \quad \mathbf{G} = -\mathbf{i} + 2\mathbf{j} - 2\mathbf{k},$$

and find the angle it makes with the x axis.

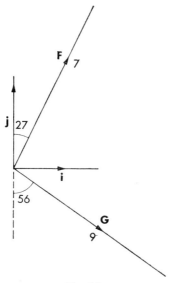

Fig. 14.

Ans. If $\mathbf{H} = \mathbf{F} + \mathbf{G}$, $\mathbf{H} = 3\mathbf{i} - 5\mathbf{j} + 8\mathbf{k} - \mathbf{i} + 2\mathbf{j} - 2\mathbf{k} = 2\mathbf{i} - 3\mathbf{j} + 6\mathbf{k}$. From eqn. (11), $H^2 = 2^2 + 3^2 + 6^2$; thus $H = 7$.

The required unit vector is therefore

$$\mathbf{H}/H = (2/7)\mathbf{i} - (3/7)\mathbf{j} + (6/7)\mathbf{k}.$$

From eqn. (10), $\theta_x = \cos^{-1}(H_x/H) = \cos^{-1}(2/7) = 73° 24'$.

(2) Calculate the magnitude and direction of the resultant of velocities 7 m.p.h. in direction N 27° E and 9 m.p.h. in direction S 56° E.

Ans. Let **i** and **j** represent velocities of 1 m.p.h. in directions *E* and *N* respectively as shown in Fig. 14. Then if **F** and **G** represent respectively velocities 7 m.p.h. in direction N 27° E and 9 m.p.h. in direction S 56° E

$$\mathbf{F} = 7 \cos 63°\mathbf{i} + 7 \cos 27°\mathbf{j} = 3{\cdot}178\mathbf{i} + 6{\cdot}237\mathbf{j},$$

$$\mathbf{G} = 9 \cos 34°\mathbf{i} - 9 \cos 56°\mathbf{j} = 7{\cdot}461\mathbf{i} - 5{\cdot}033\mathbf{j}.$$

The resultant of the two velocities is

$$\mathbf{F} + \mathbf{G} = 10{\cdot}639\mathbf{i} + 1{\cdot}204\mathbf{j},$$

and the magnitude of this is

$$\sqrt{10{\cdot}639^2 + 1{\cdot}204^2} = 10{\cdot}707 \text{ m.p.h.}$$

If θ is the angle between the resultant velocity and the *N* direction, $\tan \theta = 10{\cdot}639/1{\cdot}204$ and therefore $\theta = 83° 33'$. Thus the resultant velocity is 10·71 m.p.h. in a direction N 83° 33′ E.

Exercises

(1) Calculate the minimum velocity to be added to velocities of 4 m.p.h. in a direction N 32° E and 6 m.p.h. in a direction N 72° E to yield a resultant velocity due N.

(2) Calculate the magnitude of the resultant of a force of 7 dynes making angles 65° and 53° with the positive *x* and *z* directions respectively, and a force of 12 dynes making angles 48° and 71° respectively with the same directions; in each case the force has a positive *y* resolute. Find also the angles made by the resultant with the positive *x* and *z* directions.

(3) A boat moving at 6 m.p.h. finds the wind to be from N. At 12 m.p.h. it finds the wind to be from N.E. Calculate the magnitude and direction of the velocity of the wind.

    [*Hint.* Let the velocity of the wind be $v_x\mathbf{i} + v_y\mathbf{j}$ and obtain equations for v_x and v_y.]

(4) Prove that the five vectors $2\mathbf{i} + 3\mathbf{j} - 4\mathbf{k}, -5\mathbf{i} - \mathbf{j} + 3\mathbf{k}, 3\mathbf{i} - 2\mathbf{j} + 6\mathbf{k}, -4\mathbf{i} + 2\mathbf{j} - \mathbf{k}, 4\mathbf{i} - 2\mathbf{j} - 4\mathbf{k}$ placed successively form a closed pentagon, and calculate the lengths of its five diagonals.

2

Products of Vectors

2.1 THE SCALAR PRODUCT

We saw in the last chapter that it was necessary to define from first principles the addition of vectors, since the concept of addition derived from scalars did not lead unambiguously to any result for vectors. The same is true of the multiplication of two or more vectors, since here again our concept of multiplication of scalars does not allow of an obvious generalisation to vectors. We are thus able to define the product of vectors in any convenient way; i.e. in any way which is fruitful for the development of vectors as a tool to deal with physical quantities. It transpires that there are two such possible useful definitions. One of these, the so-called *scalar product*, will be dealt with in this section, while the other—the *vector product*—will be considered in the next section. The relevance of each of these to physical entities will then be discussed.

The scalar product of two vectors $\mathbf{p}$ and $\mathbf{q}$, inclined at angle θ as shown in Fig. 15, is defined as the product of the magnitudes of the vectors and the cosine of the angle between them. It is written $\mathbf{p} \cdot \mathbf{q}$ and referred to as "p dot q"; i.e.

$$\mathbf{p} \cdot \mathbf{q} = pq \cos \theta. \tag{16}$$

It must be emphasised that the scalar product, although derived from two vectors, is itself by definition a scalar.

Properties of the Scalar Product

It is clear from the definition (16) that $\mathbf{p.q} = \mathbf{q.p}$, and it may readily be shown by reference to Fig. 16 that

$$\mathbf{p.(q + r)} = \mathbf{p.q} + \mathbf{p.r}$$

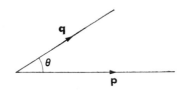

Fig. 15.

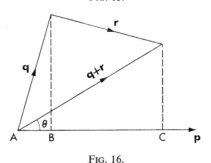

Fig. 16.

since

$$\mathbf{p.(q + r)} = p|\mathbf{q} + \mathbf{r}| \cos\theta = p \times AC = p \times AB + p \times BC$$
$$= \mathbf{p.q} + \mathbf{p.r}. \tag{17}$$

This means that products of sums of vectors can be expanded as in ordinary algebra, e.g.

$$\mathbf{(p + q).(r + s + t)} = \mathbf{p.r} + \mathbf{p.s} + \mathbf{p.t} + \mathbf{q.r} + \mathbf{q.s} + \mathbf{q.t}.$$

It is clear that if **p** and **q** are parallel, $\mathbf{p} \cdot \mathbf{q} = pq$; i.e. the scalar product is then the product of the magnitudes. In particular $\mathbf{p} \cdot \mathbf{p} = p^2$, and the left-hand side of this equation is sometimes denoted by $\mathbf{p}^2$, so that $\mathbf{p}^2 = p^2$. If **p** and **q** are perpendicular, $\cos \theta = 0$ and thus $\mathbf{p} \cdot \mathbf{q} = 0$. Conversely if it is known that $\mathbf{p} \cdot \mathbf{q} = 0$ we may conclude that *either* $\mathbf{p} = 0$ *or* $\mathbf{q} = 0$ *or* **p** and **q** are mutually perpendicular.

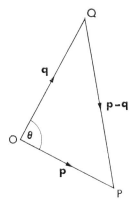

FIG. 17.

A simple illustration of the scalar product is in proving the cosine formula for a triangle. Thus, referring to Fig. 17,

$$PQ^2 = (\mathbf{p} - \mathbf{q}) \cdot (\mathbf{p} - \mathbf{q}) = \mathbf{p} \cdot \mathbf{p} - 2\mathbf{p} \cdot \mathbf{q} + \mathbf{q} \cdot \mathbf{q}$$
$$= OP^2 + OQ^2 - 2OP \cdot OQ \cos \theta.$$

Consider now all possible scalar products arising from the mutually perpendicular unit vectors **i**, **j**, **k**, discussed in §1.5. Since these all have unit magnitude and any pair are either parallel (if identical), or perpendicular, it follows immediately from eqn. (16) that

$$\mathbf{i} \cdot \mathbf{i} = \mathbf{j} \cdot \mathbf{j} = \mathbf{k} \cdot \mathbf{k} = 1, \quad \mathbf{i} \cdot \mathbf{j} = \mathbf{j} \cdot \mathbf{k} = \mathbf{i} \cdot \mathbf{k} = 0. \qquad (18)$$

Now taking the scalar product of each side of eqn. (9) with **i** yields

$$\mathbf{F} \cdot \mathbf{i} = F_x \mathbf{i} \cdot \mathbf{i} + F_y \mathbf{j} \cdot \mathbf{i} + F_z \mathbf{k} \cdot \mathbf{i},$$

and hence from eqn. (18)

$$F_x = \mathbf{F} \cdot \mathbf{i} \quad \text{and similarly} \quad F_y = \mathbf{F} \cdot \mathbf{j}, \, F_z = \mathbf{F} \cdot \mathbf{k}.$$

Thus

$$\mathbf{F} = (\mathbf{F} \cdot \mathbf{i})\mathbf{i} + (\mathbf{F} \cdot \mathbf{j})\mathbf{j} + (\mathbf{F} \cdot \mathbf{k})\mathbf{k}. \tag{19}$$

The scalar product of two vectors can now be evaluated in terms of the resolutes of the vectors. Following eqn. (9), let

$$\mathbf{p} = p_x \mathbf{i} + p_y \mathbf{j} + p_z \mathbf{k}$$

and

$$\mathbf{q} = q_x \mathbf{i} + q_y \mathbf{j} + q_z \mathbf{k}.$$

Then

$$\mathbf{p} \cdot \mathbf{q} = (p_x \mathbf{i} + p_y \mathbf{j} + p_z \mathbf{k}) \cdot (q_x \mathbf{i} + q_y \mathbf{j} + q_z \mathbf{k})$$

$$= p_x q_x \mathbf{i} \cdot \mathbf{i} + p_x q_y \mathbf{i} \cdot \mathbf{j} + p_x q_z \mathbf{i} \cdot \mathbf{k} + p_y q_x \mathbf{j} \cdot \mathbf{i} + p_y q_y \mathbf{j} \cdot \mathbf{j} +$$

$$+ p_y q_z \mathbf{j} \cdot \mathbf{k} + p_z q_x \mathbf{k} \cdot \mathbf{i} + p_z q_y \mathbf{k} \cdot \mathbf{j} + p_z q_z \mathbf{k} \cdot \mathbf{k}$$

$$= p_x q_x + p_y q_y + p_z q_z,$$

making use of the results (18).

Thus

$$\boxed{\mathbf{p} \cdot \mathbf{q} = p_x q_x + p_y q_y + p_z q_z.} \tag{20}$$

Equation (11) follows immediately from this result by letting $\mathbf{p} = \mathbf{q} = \mathbf{F}$.

Direction Cosines

For a general vector **F**, the cosines of the the angles θ_x, θ_y, θ_z which it makes with the x, y and z directions respectively are often termed the *direction cosines* of **F**, and are denoted by *l*, *m*, *n*. Thus

$$l = \cos \theta_x, \quad m = \cos \theta_y, \quad n = \cos \theta_z. \tag{21}$$

The scalar product may now be used to find the angle between the vectors **F** and **F′**, where these have direction cosines (l, m, n), (l', m', n') respectively. We have from eqns. (9), (10) and (21),

$$\mathbf{F} = F(l\mathbf{i} + m\mathbf{j} + n\mathbf{k}), \quad \mathbf{F'} = F'(l'\mathbf{i} + m'\mathbf{j} + n'\mathbf{k}).$$

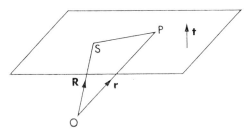

FIG. 18.

Thus $\mathbf{F}.\mathbf{F'} = FF'(ll' + mm' + nn')$ from eqn. (20), and also $\mathbf{F}.\mathbf{F'} = FF' \cos \theta$ where θ is the angle between **F** and **F′**. Hence equating these two expressions for $\mathbf{F}.\mathbf{F'}$ yields

$$\cos \theta = ll' + mm' + nn'. \tag{22}$$

The concepts discussed above may be illustrated by finding the equation of a plane passing through a given point and perpendicular to a given direction. Let S be the given point, P any point in the plane and **t** a unit vector in the given direction as shown in Fig. 18. Then if O is a fixed origin and $\overrightarrow{OS} = R$, $\overrightarrow{OP} = \mathbf{r}$, it is clear that $\overrightarrow{SP}$ ($= \mathbf{r} - \mathbf{R}$), being a vector in the plane, is perpendicular to **t**. Thus

$$(\mathbf{r} - \mathbf{R}).\mathbf{t} = 0, \tag{23}$$

and since this is true for all **r** corresponding to points in the plane, it is the equation of the plane. If P and S have cartesian coordinates (x, y, z) and (X, Y, Z) respectively, and **t** has direction cosines (l, m, n), then

$$\mathbf{r} = \mathbf{i}x + \mathbf{j}y + \mathbf{k}z, \quad \mathbf{R} = \mathbf{i}X + \mathbf{j}Y + \mathbf{k}Z,$$

and
$$\mathbf{t} = \mathbf{i}l + \mathbf{j}m + \mathbf{k}n.$$

Thus eqn. (23) becomes

$$l(x - X) + m(y - Y) + n(z - Z) = 0$$

which is the cartesian equation of the plane.

Worked examples

(1) Find the angle θ between the vectors

$$\mathbf{p} = 2\mathbf{i} - 3\mathbf{j} + \mathbf{k} \quad \text{and} \quad \mathbf{q} = 3\mathbf{i} + 4\mathbf{j} - 4\mathbf{k}.$$

Ans. Since $\mathbf{p} \cdot \mathbf{q} = pq \cos \theta$, $\theta = \cos^{-1}(\mathbf{p} \cdot \mathbf{q}/pq)$.

Now $\mathbf{p} \cdot \mathbf{q} = p_x q_x + p_y q_y + p_z q_z = 2 \times 3 - 3 \times 4 - 1 \times 4 = -10$ and

$$pq = \sqrt{(p_x{}^2 + p_y{}^2 + p_z{}^2)(q_x{}^2 + q_y{}^2 + q_z{}^2)}$$

$$= \sqrt{(4 + 9 + 1)(9 + 16 + 16)} = \sqrt{574} = 23 \cdot 96.$$

Hence $\theta = \cos^{-1}[-(10/23 \cdot 96)] = \cos^{-1}(-0 \cdot 4174) = 114° \ 40'$.

(2) For what values of a are the vectors $2\mathbf{i} + a\mathbf{j} + 3\mathbf{k}$, $a\mathbf{i} - a\mathbf{j} + \mathbf{k}$ perpendicular?

Ans. For the vectors to be perpendicular, their scalar product is zero; i.e., $2a - a^2 + 3 = 0$ and therefore $a = -1$ or 3.

(3) Prove that in a tetrahedron, if two pairs of opposite edges are perpendicular, then the third pair is also perpendicular and the sums of squares on opposite edges is the same for each pair.

Ans. Using the notation of Fig. 8 (p. 8), $\overrightarrow{BA} = \mathbf{r} - \mathbf{s}$, $\overrightarrow{CB} = \mathbf{s} - \mathbf{t}$ and $\overrightarrow{AC} = \mathbf{t} - \mathbf{r}$. Then, since $OA \perp BC$, $\mathbf{r}.(\mathbf{s} - \mathbf{t}) = 0$, and thus $\mathbf{r}.\mathbf{s} = \mathbf{r}.\mathbf{t}$. Similarly, since $OB \perp AC$, $\mathbf{s}.\mathbf{t} = \mathbf{s}.\mathbf{r}$, and hence $\mathbf{r}.\mathbf{s} = \mathbf{s}.\mathbf{t} = \mathbf{t}.\mathbf{r}$. Therefore, $\mathbf{t}.(\mathbf{r} - \mathbf{s}) = 0$, and so $OC \perp AB$. Further, $OA^2 + BC^2 = \mathbf{r}^2 + (\mathbf{s} - \mathbf{t})^2 = \mathbf{r}^2 + \mathbf{s}^2 + \mathbf{t}^2 - 2\mathbf{s}.\mathbf{t}$. Now, since the first three terms are symmetric in $\mathbf{r}$, $\mathbf{s}$ and $\mathbf{t}$, they will be the same for any other pair of edges, and the fourth term will also be the same, as proved above. Hence the sum of the squares will be the same for each pair of edges.

Exercises

(1) Evaluate the scalar product of the two vectors $2\mathbf{i} - 3\mathbf{j} + 5\mathbf{k}$, $3\mathbf{i} + \mathbf{j} - 2\mathbf{k}$ and hence find the angle between them.

(2) Two adjacent sides of a parallelogram are the vectors $\mathbf{i} + 2\mathbf{j} + 3\mathbf{k}$ and $-2\mathbf{i} + \mathbf{j} - 2\mathbf{k}$. Find the angle between the two diagonals of the parallelogram, and the angles which the shorter diagonal makes with each side.

(3) The vectors $\overrightarrow{AB}$ and $\overrightarrow{AC}$ make angles 49°, 72° and 111°, 68° with the positive x and z directions respectively and both have a positive resolute in the y direction. Calculate the angles of the triangle ABC.

(4) By applying eqn. (22) to vectors lying in the x–y plane, prove the formula $\cos(A - B) = \cos A \cos B + \sin A \sin B$.

2.2 THE VECTOR PRODUCT

The *vector product* of two vectors $\mathbf{p}$ and $\mathbf{q}$ inclined at angle θ is defined as a vector of magnitude equal to the product of the magnitudes of the given vectors and the sine of the angle between them. Its direction is perpendicular to the plane of $\mathbf{p}$ and $\mathbf{q}$, in the direction of motion of a right-handed screw rotated along the shortest path from $\mathbf{p}$ to $\mathbf{q}$. It is written $\mathbf{p} \wedge \mathbf{q}$ and referred to as " p cross q ". Thus

$$\mathbf{p} \wedge \mathbf{q} = pq \sin \theta \mathbf{n}, \tag{24}$$

where $\mathbf{n}$ is a unit vector perpendicular to the plane of $\mathbf{p}$ and $\mathbf{q}$ in the direction of motion of a right-handed screw rotating from $\mathbf{p}$ to $\mathbf{q}$ along the shortest path, as shown in Fig. 19. It should again be emphasised that *by definition* the vector product of two vectors is itself a vector, in contradistinction to the scalar product which is a scalar.

A rather unexpected result follows immediately from the definition (24) if we consider $\mathbf{q} \wedge \mathbf{p} = qp \sin \theta \mathbf{m}$. Now, the first three terms in this product are identical with the corresponding terms in the expression for $\mathbf{p} \wedge \mathbf{q}$. The vector $\mathbf{m}$,

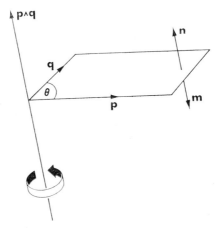

Fig. 19.

however, is a unit vector perpendicular to the plane of $\mathbf{p}$ and $\mathbf{q}$, in the direction of motion of a right-handed screw rotating *from* $\mathbf{q}$ *to* $\mathbf{p}$ along the shortest path. Clearly the vector $\mathbf{m}$, arising from a rotation *from* $\mathbf{q}$ *to* $\mathbf{p}$ is in the *opposite* direction to the vector $\mathbf{n}$ arising from a rotation *from* $\mathbf{p}$ *to* $\mathbf{q}$, i.e. $\mathbf{m} = -\mathbf{n}$, and thus

$$\mathbf{q} \wedge \mathbf{p} = -\mathbf{p} \wedge \mathbf{q}. \tag{25}$$

This means that unlike ordinary products, the vector product of two vectors does *not commute*; i.e. $\mathbf{p} \wedge \mathbf{q} \neq \mathbf{q} \wedge \mathbf{p}$. It is therefore necessary to be careful to maintain the correct order of the factors in a vector product.

It may be shown that

$$p \wedge (q + r) = p \wedge q + p \wedge r, \tag{26}$$

although the proof (which is somewhat more involved than the corresponding statement for a scalar product) will not be given here. This means that vector products of vectors can be expanded in the same way as in ordinary algebra, as long as care is taken to maintain the correct order of the factors in a product. Thus, for example,

$$(p + q) \wedge (r + s) = p \wedge r + p \wedge s + q \wedge r + q \wedge s.$$

Geometrical Interpretation of Vector Product

It follows from the definition (24) together with Fig. 19 that

$$|p \wedge q| = pq \sin \theta = \text{area of parallelogram } OABC.$$

Thus the magnitude of the vector product is the area of a parallelogram with adjacent sides equal to the two vectors. Alternatively, if we define the *vector area* of a plane area as a vector with magnitude equal to the area, and direction perpendicular to the plane of the area, it follows that the vector product of two vectors is the vector area of the parallelogram formed with the two vectors as adjacent sides.

Properties of the Vector Product

It is clear that if p and q are parallel vectors, $p \wedge q = 0$ since $\theta = 0$, and in particular, $p \wedge p = 0$. Conversely, if it is known that $p \wedge q = 0$, we can conclude that *either* $p = 0$ *or*

$\mathbf{q} = 0$ *or* $\mathbf{p}$ and $\mathbf{q}$ are parallel. If $\mathbf{p}$ and $\mathbf{q}$ are perpendicular, then $|\mathbf{p} \wedge \mathbf{q}| = pq$ since $\theta = \pi/2$, and $\mathbf{p}$, $\mathbf{q}$, $\mathbf{p} \wedge \mathbf{q}$ form a set of three mutually perpendicular vectors.

Consider now all possible vector products arising from the three mutually perpendicular unit vectors $\mathbf{i}$, $\mathbf{j}$, $\mathbf{k}$ shown in Fig. 20. Clearly

$$\mathbf{i} \wedge \mathbf{i} = \mathbf{j} \wedge \mathbf{j} = \mathbf{k} \wedge \mathbf{k} = 0. \tag{27}$$

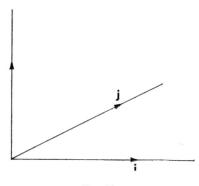

FIG. 20.

Also, since any other pair are mutually perpendicular, and each has unit magnitude,

$$\left. \begin{array}{l} \mathbf{i} \wedge \mathbf{j} = -\mathbf{j} \wedge \mathbf{i} = \mathbf{k}, \\ \mathbf{j} \wedge \mathbf{k} = -\mathbf{k} \wedge \mathbf{j} = \mathbf{i}, \\ \mathbf{k} \wedge \mathbf{i} = -\mathbf{i} \wedge \mathbf{k} = \mathbf{j}. \end{array} \right\} \tag{28}$$

The vector product of two vectors can now be evaluated in terms of the resolutes of the two vectors. Let

$$\mathbf{p} = p_x \mathbf{i} + p_y \mathbf{j} + p_z \mathbf{k} \quad \text{and} \quad \mathbf{q} = q_x \mathbf{i} + q_y \mathbf{j} + q_z \mathbf{k}$$

Then,

$$\mathbf{p} \wedge \mathbf{q} = (p_x\mathbf{i} + p_y\mathbf{j} + p_z\mathbf{k}) \wedge (q_x\mathbf{i} + q_y\mathbf{j} + q_z\mathbf{k})$$

$$= p_xq_x\mathbf{i} \wedge \mathbf{i} + p_xq_y\mathbf{i} \wedge \mathbf{j} + p_xq_z\mathbf{i} \wedge \mathbf{k} + p_yq_x\mathbf{j} \wedge \mathbf{i} +$$

$$+ p_yq_y\mathbf{j} \wedge \mathbf{j} + p_yq_z\mathbf{j} \wedge \mathbf{k} + p_zq_x\mathbf{k} \wedge \mathbf{i} + p_zq_y\mathbf{k} \wedge \mathbf{j} +$$

$$+ p_zq_z\mathbf{k} \wedge \mathbf{k}$$

$$= p_xq_y\mathbf{k} - p_xq_z\mathbf{j} - p_yq_x\mathbf{k} + p_yq_z\mathbf{i} + p_zq_x\mathbf{j} - p_zq_y\mathbf{i}.$$

Therefore

$$\boxed{\mathbf{p} \wedge \mathbf{q} = \mathbf{i}(p_yq_z - p_zq_y) + \mathbf{j}(p_zq_x - p_xq_z) + \mathbf{k}(p_xq_y - p_yq_x)}$$

$$(29)$$

making use of the above results (27) and (28). If we employ determinants, this result can be written

$$\mathbf{p} \wedge \mathbf{q} = \begin{vmatrix} \mathbf{i} & \mathbf{j} & \mathbf{k} \\ p_x & p_y & p_z \\ q_x & q_y & q_z \end{vmatrix}. \tag{30}$$

Worked examples

(1) Find a unit vector perpendicular to the plane of $2\mathbf{i} - \mathbf{j} + \mathbf{k}, 3\mathbf{i} + 4\mathbf{j} - \mathbf{k}$.

Ans. The vector product of the given vectors will be a vector perpendicular to their plane, from which a unit vector can be readily derived. Now, from the result (29)

$$(2\mathbf{i} - \mathbf{j} + \mathbf{k}) \wedge (3\mathbf{i} + 4\mathbf{j} - \mathbf{k}) = \mathbf{i}[(-1 \times -1) - (1 \times 4)] +$$

$$+ \mathbf{j}[(1 \times 3) - (2 \times -1)] + \mathbf{k}[(2 \times 4) - (-1 \times 3)]$$

$$= -3\mathbf{i} + 5\mathbf{j} + 11\mathbf{k}.$$

The magnitude of the vector product is

$$(3^2 + 5^2 + 11^2)^{\frac{1}{2}} = \sqrt{155},$$

and thus the required unit vector is

$$155^{-\frac{1}{2}}(-3\mathbf{i} + 5\mathbf{j} + 11\mathbf{k}).$$

(2) Find the area of the triangle PQR, the cartesian co-ordinates of whose vertices are $P(1, 3, 4)$, $Q(-2, 1, -1)$, $R(0, -3, 2)$.

Ans. If $\mathbf{i}, \mathbf{j}, \mathbf{k}$ represent unit distances in the x, y, z directions, $\overrightarrow{PQ} = -3\mathbf{i} - 2\mathbf{j} - 5\mathbf{k}$ and $\overrightarrow{PR} = -\mathbf{i} - 6\mathbf{j} - 2\mathbf{k}$. From the geometrical significance of the vector product, given earlier, it is clear that

$$\text{area of } \triangle PQR = \tfrac{1}{2}\left|\overrightarrow{PQ} \wedge \overrightarrow{PR}\right|.$$

Now,

$$\overrightarrow{PQ} \wedge \overrightarrow{PR} = \mathbf{i}[(-2\times -2) - (-5\times -6)] +$$

$$+ \mathbf{j}[(-5\times -1) - (-3\times -2)] + \mathbf{k}[(-3\times -6) - (-1\times -2)]$$

$$= -26\mathbf{i} - \mathbf{j} + 16\mathbf{k}.$$

Thus area of $\triangle PQR = \tfrac{1}{2}(26^2 + 1 + 16^2)^{\frac{1}{2}} = \tfrac{1}{2}\sqrt{933}$.

Exercises

(1) Prove $(\mathbf{p.q})^2 + |\mathbf{p} \wedge \mathbf{q}|^2 = (\mathbf{p.p})(\mathbf{q.q})$.

(2) If $p = \mathbf{i} + 2\mathbf{j} - 3\mathbf{k}$, $q = -2\mathbf{i} + 4\mathbf{j} + \mathbf{k}$, $r = 2\mathbf{i} + \mathbf{j} + 3\mathbf{k}$, evaluate $\mathbf{p} \wedge \mathbf{q}$, $\mathbf{p} \wedge \mathbf{r}$, $\mathbf{q} \wedge \mathbf{r}$ and $(\mathbf{p} - 2\mathbf{q} + 3\mathbf{r}) \wedge (3\mathbf{p} + 4\mathbf{q} - 2\mathbf{r})$. Hence find a vector of magnitude 3 perpendicular to the plane of

$$\mathbf{p} - 2\mathbf{q} + 3\mathbf{r} \quad \text{and} \quad 3\mathbf{p} + 4\mathbf{q} - 2\mathbf{r}.$$

(3) Using the relevant vector product, calculate the area of the triangle *ABC* given in §2.1, example 3.

(4) Prove (i) $(\mathbf{p} + \mathbf{q}).(\mathbf{p} - \mathbf{q}) \equiv p^2 - q^2$.
   (ii) $(\mathbf{p} + \mathbf{q}) \wedge (\mathbf{p} - \mathbf{q}) \equiv 2\mathbf{q} \wedge \mathbf{p}$.

If **p** and **q** are adjacent sides of a parallelogram, what is the geo-metrical interpretation of these two identities?

2.3 APPLICATIONS OF SCALAR AND VECTOR PRODUCTS

Work

The work done by a force acting on a particle is defined in elementary dynamics as the product of the force and the distance moved by its point of application in the direction of the force. It is a scalar quantity bearing no notion of direction. Now, suppose the force to be represented by a vector **F**, and suppose that the actual displacement of the point of

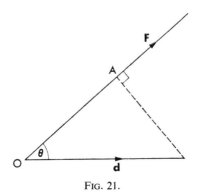

FIG. 21.

application is represented by a vector **d** at angle θ to **F** as shown in Fig. 21, then clearly the work W is given by

$$W = F \times OA = F d \cos \theta = \mathbf{F} . \mathbf{d}; \tag{31}$$

i.e. the work done is the scalar product of the force and displacement vectors. Making use of this result, it may easily be proved that the work done by two forces **F** and **G** is equal to the work done by their resultant **R**. For the total work done by **F** and **G** is

$$\mathbf{F} . \mathbf{d} + \mathbf{G} . \mathbf{d} = (\mathbf{F} + \mathbf{G}) . \mathbf{d} = \mathbf{R} . \mathbf{d}, \tag{32}$$

since $\mathbf{R} = \mathbf{F} + \mathbf{G}$, and the R.H.S. of eqn. (32) is the work done by $\mathbf{R}$.

Moment of a Force

In elementary two-dimensional work, the moment of a force $\mathbf{F}$ about a point O is defined as the product of the

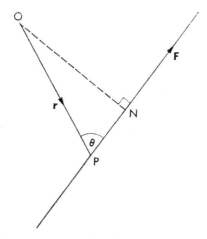

Fɪɢ. 22.

magnitude of the force and the perpendicular distance of the point from the line of action of the force. Referring to Fig. 22, the moment of $\mathbf{F}$ about O is therefore $F \times ON$. In three dimensions, however, the definition of moment of a force must be generalised in two ways. First, we define the moment of $\mathbf{F}$ about an *axis* (*not* a point) as the product of F and the perpendicular distance of the axis from the line of action of $\mathbf{F}$. The result derived above from Fig. 22 will thus remain unaltered if the axis chosen is perpendicular to the plane of the paper as well as passing through O. Secondly, the moment is defined as a *vector* (*not* a scalar as in the elementary approach) whose magnitude is $F \times ON$ and whose direction is perpendicular to the plane containing $\mathbf{F}$ and O.

If P is a point on the line of action of $\mathbf{F}$ and $\overrightarrow{OP} = \mathbf{r}$, this complete definition gives the moment $\mathbf{M}$ of $\mathbf{F}$ about the axis through O as

$$\mathbf{M} = \mathbf{r} \wedge \mathbf{F}, \tag{33}$$

since $|\mathbf{r} \wedge \mathbf{F}| = Fr \sin \theta = F \times ON$, and the direction of $\mathbf{r} \wedge \mathbf{F}$ is perpendicular to the plane of $\mathbf{r}$ and $\mathbf{F}$ as required. Equation (33) may be readily used to show that the sum of the moments of two forces is equal to the moment of the resultant; the proof follows the same lines as that given above for work. It should be noted that the reason that the " scalar definition " of moment is sufficient in two dimensions is because the axis is always perpendicular to the plane of the forces, and thus the corresponding vector moments are all parallel to this axis and may therefore be added in scalar fashion. In three dimensions, however, the vector moments are not necessarily parallel to any fixed direction, and must therefore always be added vectorially.

Angular Momentum

The angular momentum $\mathbf{P}$ of a point mass m with velocity $\mathbf{v}$ about an axis passing through O is defined as the moment of momentum of m about O; i.e. since the momentum of m is $m\mathbf{v}$,

$$\mathbf{P} = \mathbf{r} \wedge m\mathbf{v} \tag{34}$$

in the notation of Fig. 23. We may employ this definition to obtain a simplified expression for the angular momentum of a system of particles. Suppose we have a set of N particles of masses $m_1, m_2, \ldots, m_N$ at points with position vectors $\mathbf{r}_1, \mathbf{r}_2, \ldots, \mathbf{r}_N$ relative to O and moving with velocities $\mathbf{v}_1$,

$\mathbf{v}_2, \ldots, \mathbf{v}_N$. Then the vector position $\mathbf{R}$ of the centre of gravity G of the system is defined by

$$(m_1 + m_2 + \ldots m_N)\mathbf{R} = m_1\mathbf{r}_1 + m_2\mathbf{r}_2 + \ldots m_N\mathbf{r}_N, \dagger$$

(35)

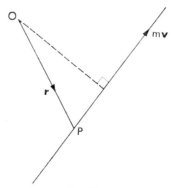

FIG. 23.

and this implies, as will be shown in Chapter 4, that

$$(m_1 + m_2 + \ldots m_N)\mathbf{V} = m_1\mathbf{v}_1 + m_2\mathbf{v}_2 + \ldots m_N\mathbf{v}_N, \quad (36)$$

where $\mathbf{V}$ is the velocity of the c.g. Now referring to Fig. 24 we may write $\mathbf{r}_p = \mathbf{R} + \mathbf{S}_p$ for any mass m_p, where $\mathbf{S}_p$ is its vector position relative to the c.g., and $\mathbf{v}_p = \mathbf{V} + \mathbf{W}_p$, where $\mathbf{W}_p$ is its velocity relative to the c.g. From eqn. (34) it is seen that the total angular momentum of the system is given by

$$\mathbf{P} = \sum_{p=1}^{N} \mathbf{r}_p \wedge (m_p\mathbf{v}_p) = \sum_{p=1}^{N} (\mathbf{R} + \mathbf{S}_p) \wedge m_p(\mathbf{V} + \mathbf{W}_p)$$

$$= (\sum_{p=1}^{N} m_p)\mathbf{R} \wedge \mathbf{V} + \mathbf{R} \wedge (\sum_{p=1}^{N} m_p\mathbf{W}_p) + (\sum_{p=1}^{N} m_p\mathbf{S}_p) \wedge \mathbf{V} +$$

$$+ \sum_{p=1}^{N} \mathbf{S}_p \wedge m_p\mathbf{W}_p, \tag{37}$$

† It may readily be shown that this agrees with the simpler scalar definition usually given for mass distributions in one or two dimensions.

where the factors **R** and **V**, being independent of p, have been taken out of the summations. Now, since the vectors $\mathbf{S}_p$ are all taken relative to G as origin, it follows from eqn. (35) that

$$\sum_{p=1}^{N} m_p \mathbf{S}_p = 0$$

since G is the centre of gravity, and similarly

$$\sum_{p=1}^{N} m_p \mathbf{W}_p = 0$$

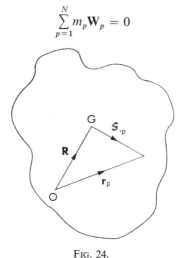

Fig. 24.

from eqn. (36). Thus the second and third terms in eqn. (37) are zero, and hence

$$\mathbf{P} = \mathbf{R} \wedge (M\mathbf{V}) + \sum_{p=1}^{N} \mathbf{S}_p \wedge m\mathbf{W}_p, \tag{38}$$

where $M = \sum_{p=1}^{N} m_p$ is the total mass of the system. The first term in eqn. (38) clearly depends only on the motion of the c.g., while the second depends only on the motion of the system relative to the c.g. This final expression (38) is used considerably in treating the dynamics of systems of particles.

Angular Velocity

Consider a rigid body rotating with angular velocity ω about the axis OZ, shown in Fig. 25, and take a point P on the body such that $\overrightarrow{OP} = \mathbf{r}$. What is the velocity of P? Now if PN is perpendicular to OZ, the velocity $\mathbf{v}$ of P will be perpendicular to the plane OPN and its magnitude will be

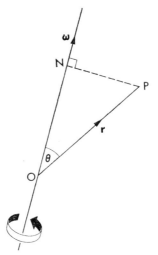

Fig. 25.

$\omega PN = \omega r \sin \theta$. Therefore, if we define the vector angular velocity $\boldsymbol{\omega}$ of the body as a vector of magnitude ω with direction along the axis of rotation, in the direction of motion of a right-handed screw rotating with the body, we have

$$\mathbf{v} = \boldsymbol{\omega} \wedge \mathbf{r}, \tag{39}$$

since eqn. (39) yields the correct magnitude and direction for $\mathbf{v}$. It follows immediately that if the body is subject simultaneously to two angular velocities $\boldsymbol{\omega}_1$ and $\boldsymbol{\omega}_2$, each about an axis passing through O, then the resultant velocity of P is given by $(\omega_1 + \omega_2) \wedge \mathbf{r}$.

Exercises

(1) What is the work done when a force of 7 dynes parallel to the vector $2\mathbf{i} - 3\mathbf{j} + \mathbf{k}$ moves its point of application from $P(1, 2, 3)$ to $Q(-2, 4, 1)$; a unit on the coordinate axes is 1 cm?

(2) A rigid body is spinning with simultaneous angular velocities of 3 radians per sec parallel to $\mathbf{i} + \mathbf{j} - \mathbf{k}$, and 4 radians per sec parallel to $2\mathbf{i} + 3\mathbf{j} + \mathbf{k}$ about axes which intersect at D with position vector $-2\mathbf{i} - \mathbf{j} + 3\mathbf{k}$. What are the possible speeds of the point P with position vector $2\mathbf{i} - 2\mathbf{j} + \mathbf{k}$?

3

Products of Three or Four Vectors

3.1 THE SCALAR TRIPLE PRODUCT

In the last chapter we discussed the scalar and vector products of two vectors. What products of three vectors can now be defined making use of these two types of product? If we consider $\mathbf{q}.\mathbf{r}$, then this being a scalar can only be combined with a third vector $\mathbf{p}$ in the form $(\mathbf{q}.\mathbf{r})\mathbf{p}$ which is, of course, a vector parallel to $\mathbf{p}$ and with magnitude increased by a factor $\mathbf{q}.\mathbf{r}$. Taking now the product $\mathbf{q} \wedge \mathbf{r}$, this being a vector can be combined with a third vector $\mathbf{p}$ in the two forms $\mathbf{p}.(\mathbf{q} \wedge \mathbf{r})$ and $\mathbf{p} \wedge (\mathbf{q} \wedge \mathbf{r})$. The first of these forms is known as the scalar triple product and will be considered in the present section, while the second form, known as the vector triple product, will be dealt with in the next section.

The scalar triple product $\mathbf{p}.(\mathbf{q} \wedge \mathbf{r})$ can be given a simple geometrical interpretation by consideration of a parallelepiped with edges $\mathbf{p}$, $\mathbf{q}$, $\mathbf{r}$ as shown in Fig. 26. Let $\mathbf{n}$ be a unit vector perpendicular to the parallelogram $ABCD$, and let h be the perpendicular distance of E from $ABCD$. Then if S is the area of $ABCD$, the volume V of the parallelepiped is given by

$$V = h \times S = (\mathbf{p}.\mathbf{n})|\mathbf{q} \wedge \mathbf{r}|$$

$$= \mathbf{p}.[|\mathbf{q} \wedge \mathbf{r}|\mathbf{n}] = \mathbf{p}.(\mathbf{q} \wedge \mathbf{r}); \tag{40}$$

i.e. $\mathbf{p}.(\mathbf{q} \wedge \mathbf{r})$ is the volume of a parallelepiped with edges $\mathbf{p}$, $\mathbf{q}$, $\mathbf{r}$. It is clear that any other face, e.g. *ABFE* or *ADHE* could have been taken as the base of the parallelepiped, and this would have given $V = \mathbf{r}.(\mathbf{p} \wedge \mathbf{q})$ or $\mathbf{q}.(\mathbf{r} \wedge \mathbf{p})$. Hence it follows that

$$\mathbf{p}.(\mathbf{q} \wedge \mathbf{r}) = \mathbf{q}.(\mathbf{r} \wedge \mathbf{p}) = \mathbf{r}.(\mathbf{p} \wedge \mathbf{q}), \tag{41}$$

since each equals V. Thus as long as the cyclic order is maintained, the scalar triple product is independent of the

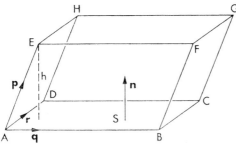

FIG. 26.

position of the dot and cross products occurring in it, while if two of the factors are exchanged the product changes sign. It is clear from this geometrical interpretation of the scalar triple product that $\mathbf{p}.(\mathbf{q} \wedge \mathbf{r}) = 0$ if $\mathbf{p}$, $\mathbf{q}$ and $\mathbf{r}$ are coplanar, since then $V = 0$, and in particular it is zero if two of the vectors are identical.

Making use of eqns. (20) and (29), $\mathbf{p}.(\mathbf{q} \wedge \mathbf{r})$ may be readily evaluated in terms of the resolutes of the three vectors $\mathbf{p}$, $\mathbf{q}$ and $\mathbf{r}$ to yield

$$\mathbf{p}.(\mathbf{q} \wedge \mathbf{r}) = p_x(q_y r_z - q_z r_y) + p_y(q_z r_x - q_x r_z) + p_z(q_x r_y - q_y r_x) \tag{42}$$

$$= \begin{vmatrix} p_x & p_y & p_z \\ q_x & q_y & q_z \\ r_x & r_y & r_z \end{vmatrix}. \tag{42a}$$

The invariance of this determinant for cyclic interchange of rows corresponds to the invariance of $\mathbf{p}.(\mathbf{q} \wedge \mathbf{r})$ for cyclic interchange of factors.

The vector equation of a plane passing through points A, B, C such that $\overrightarrow{OA} = \mathbf{r}_1$, $\overrightarrow{OB} = \mathbf{r}_2$, $\overrightarrow{OC} = \mathbf{r}_3$, as shown in Fig. 27 may now be obtained as follows. For any point P in the plane such that $\overrightarrow{OP} = \mathbf{r}$, $\overrightarrow{AP}(= \mathbf{r} - \mathbf{r}_1)$ and $\overrightarrow{BP}(= \mathbf{r} - \mathbf{r}_2)$

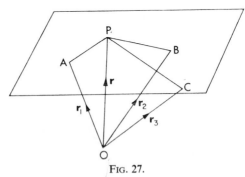

FIG. 27.

and $\overrightarrow{CP} = (\mathbf{r} - \mathbf{r}_3)$ are coplanar. Hence $\overrightarrow{AP}.(\overrightarrow{BP} \wedge \overrightarrow{CP}) = 0$ and thus the equation of the plane is

$$(\mathbf{r} - \mathbf{r}_1).[(\mathbf{r} - \mathbf{r}_2) \wedge (\mathbf{r} - \mathbf{r}_3)] = 0, \tag{43}$$

which may be expanded to give

$$\mathbf{r}.[(\mathbf{r}_1 \wedge \mathbf{r}_2) + (\mathbf{r}_2 \wedge \mathbf{r}_3) + (\mathbf{r}_3 \wedge \mathbf{r}_1)] = \mathbf{r}_1.(\mathbf{r}_2 \wedge \mathbf{r}_3). \tag{44}$$

If this is written in terms of cartesian components, it is seen from eqn. (42a) that it becomes

$$\begin{vmatrix} x & y & z \\ x_1 & y_1 & z_1 \\ x_2 & y_2 & z_2 \end{vmatrix} + \begin{vmatrix} x & y & z \\ x_2 & y_2 & z_2 \\ x_3 & y_3 & z_3 \end{vmatrix} + \begin{vmatrix} x & y & z \\ x_3 & y_3 & z_3 \\ x_1 & y_1 & z_1 \end{vmatrix} = \begin{vmatrix} x_1 & y_1 & z_1 \\ x_2 & y_2 & z_2 \\ x_3 & y_3 & z_3 \end{vmatrix}$$

since $(\mathbf{r}_1)_x = x_1$, etc., and this is the cartesian equation for such a plane.

Exercise

If $\overrightarrow{OP} = \mathbf{p}$, $\overrightarrow{OQ} = \mathbf{q}$, $\overrightarrow{OR} = \mathbf{r}$ and $\overrightarrow{OS} = \mathbf{s}$, show that the volume of the tetrahedron $PQRS$ is

$$\tfrac{1}{6}\{[\mathbf{p}.(\mathbf{q}\wedge\mathbf{r})] - [\mathbf{p}.(\mathbf{q}\wedge\mathbf{s})] + [\mathbf{p}.(\mathbf{r}\wedge\mathbf{s})] - [\mathbf{q}.(\mathbf{r}\wedge\mathbf{s})]\}.$$

3.2   THE VECTOR TRIPLE PRODUCT

We proceed to prove that

$$\mathbf{p}\wedge(\mathbf{q}\wedge\mathbf{r}) \equiv (\mathbf{p}.\mathbf{r})\mathbf{q} - (\mathbf{p}.\mathbf{q})\mathbf{r}. \tag{45}$$

Since $\mathbf{q}\wedge\mathbf{r}$ is perpendicular to the plane of $\mathbf{q}$ and $\mathbf{r}$, it is clear that $\mathbf{p}\wedge(\mathbf{q}\wedge\mathbf{r})$ will lie in the plane of $\mathbf{q}$ and $\mathbf{r}$, and therefore equals $a\mathbf{q} + b\mathbf{r}$ for some a and b. To show that $a = \mathbf{p}.\mathbf{r}$ and $b = \mathbf{p}.\mathbf{q}$, we choose cartesian axes with OX parallel to $\mathbf{q}$, OY in the plane of $\mathbf{q}$ and $\mathbf{r}$, and OZ parallel to $\mathbf{q}\wedge\mathbf{r}$. Then

$$\mathbf{q} = q_x\mathbf{i}, \quad \mathbf{r} = r_x\mathbf{i} + r_y\mathbf{j} \quad \text{and} \quad \mathbf{p} = p_x\mathbf{i} + p_y\mathbf{j} + p_z\mathbf{k}. \quad \text{Thus}$$

$\mathbf{q}\wedge\mathbf{r} = q_x r_y\mathbf{k}$ from eqn. (29), and hence

$$\mathbf{p}\wedge(\mathbf{q}\wedge\mathbf{r}) = p_y q_x r_y\mathbf{i} - p_x q_x r_y\mathbf{j}. \tag{46}$$

Also $\mathbf{p}.\mathbf{r} = p_x r_x + p_y r_y$ and $\mathbf{p}.\mathbf{q} = p_x q_x$, so that

$$(\mathbf{p}.\mathbf{r})\mathbf{q} - (\mathbf{p}.\mathbf{q})\mathbf{r} = p_x r_x q_x\mathbf{i} + p_y r_y q_x\mathbf{i} - p_x q_x r_x\mathbf{i} - p_x q_x r_y\mathbf{j}$$

$$= p_y r_y q_x\mathbf{i} - p_x q_x r_y\mathbf{j}$$

$$= \mathbf{p}\wedge(\mathbf{q}\wedge\mathbf{r}) \text{ from eqn. (46)}.$$

The position of the bracket in $\mathbf{p}\wedge(\mathbf{q}\wedge\mathbf{r})$ is important, since from eqn. (45)

$$(\mathbf{p}\wedge\mathbf{q})\wedge\mathbf{r} = \mathbf{r}\wedge(\mathbf{q}\wedge\mathbf{p}) = (\mathbf{r}.\mathbf{p})\mathbf{q} - (\mathbf{r}.\mathbf{q})\mathbf{p},$$

which is *not* equal to the corresponding expression for $\mathbf{p}\wedge(\mathbf{q}\wedge\mathbf{r})$. In particular, $\mathbf{p}\wedge(\mathbf{p}\wedge\mathbf{q}) = (\mathbf{p}.\mathbf{q})\mathbf{p} - p^2\mathbf{q}$, while $(\mathbf{p}\wedge\mathbf{p})\wedge\mathbf{q} = 0$, since $\mathbf{p}\wedge\mathbf{p} = 0$.

Exercises

(1) Check the identity (45) by evaluating both sides for the case of $\mathbf{p} = \mathbf{i} - \mathbf{j} + \mathbf{k}$, $\mathbf{q} = 2\mathbf{i} - 3\mathbf{j} + 2\mathbf{k}$, $\mathbf{r} = 4\mathbf{i} + \mathbf{j} - 3\mathbf{k}$.

(2) Prove that for all $\mathbf{p}$, $\mathbf{q}$, $\mathbf{r}$
$$\mathbf{p} \wedge (\mathbf{q} \wedge \mathbf{r}) + \mathbf{q} \wedge (\mathbf{r} \wedge \mathbf{p}) + \mathbf{r} \wedge (\mathbf{p} \wedge \mathbf{q}) \equiv 0.$$

(3) By use of eqn. (44), or otherwise, show that if P, Q, R are non-collinear points, then for any point O, $\overrightarrow{OP} \wedge \overrightarrow{OQ} + \overrightarrow{OQ} \wedge \overrightarrow{OR} + \overrightarrow{OR} \wedge \overrightarrow{OP}$ is a vector perpendicular to the plane of P, Q, R.

(4) If a vector $\mathbf{r}$ is resolved into components parallel and perpendicular to another vector $\mathbf{q}$, show that the component perpendicular to $\mathbf{q}$ is
$$\mathbf{q} \wedge (\mathbf{r} \wedge \mathbf{q})/q^2.$$

3.3 PRODUCTS OF FOUR VECTORS

The following products involving four vectors are possible:

(a) $\mathbf{p} . [\mathbf{q} \wedge (\mathbf{r} \wedge \mathbf{s})]$,   (c) $(\mathbf{p} \wedge \mathbf{q}) . (\mathbf{r} \wedge \mathbf{s})$,

(b) $\mathbf{p} \wedge [\mathbf{q} \wedge (\mathbf{r} \wedge \mathbf{s})]$,   (d) $(\mathbf{p} \wedge \mathbf{q}) \wedge (\mathbf{r} \wedge \mathbf{s})$.

(a) can be simplified by use of eqn. (45) to yield
$$\mathbf{p} . [\mathbf{q} \wedge (\mathbf{r} \wedge \mathbf{s})] = \mathbf{p} . [(\mathbf{q}.\mathbf{s})\mathbf{r} - (\mathbf{q}.\mathbf{r})\mathbf{s}]$$
$$= (\mathbf{p}.\mathbf{r})(\mathbf{q}.\mathbf{s}) - (\mathbf{p}.\mathbf{s})(\mathbf{q}.\mathbf{r}). \tag{47}$$

Hence it follows that
$$\mathbf{p} . [\mathbf{q} \wedge (\mathbf{r} \wedge \mathbf{s})] = \mathbf{r} . [\mathbf{s} \wedge (\mathbf{p} \wedge \mathbf{q})] \tag{47a}$$

since the expansion of $\mathbf{r} . [\mathbf{s} \wedge (\mathbf{p} \wedge \mathbf{q})]$ yields the R.H.S. of eqn. (47).

(b) gives
$$\mathbf{p} \wedge [\mathbf{q} \wedge (\mathbf{r} \wedge \mathbf{s})] = \mathbf{p} \wedge [(\mathbf{q}.\mathbf{s})\mathbf{r} - (\mathbf{q}.\mathbf{r})\mathbf{s}]$$
$$= (\mathbf{q}.\mathbf{s})(\mathbf{p} \wedge \mathbf{r}) - (\mathbf{q}.\mathbf{r})(\mathbf{p} \wedge \mathbf{s}). \tag{48}$$

Alternatively,
$$\mathbf{p} \wedge [\mathbf{q} \wedge (\mathbf{r} \wedge \mathbf{s})] = [\mathbf{p}.(\mathbf{r} \wedge \mathbf{s})]\mathbf{q} - (\mathbf{p}.\mathbf{q})(\mathbf{r} \wedge \mathbf{s}), \tag{49}$$

considering $\mathbf{r} \wedge \mathbf{s}$ as a single vector. Equating (48) and (49) yields
$$(\mathbf{q}.\mathbf{s})(\mathbf{p} \wedge \mathbf{r}) + (\mathbf{q}.\mathbf{p})(\mathbf{r} \wedge \mathbf{s}) + (\mathbf{q}.\mathbf{r})(\mathbf{s} \wedge \mathbf{p}) = [\mathbf{p}.(\mathbf{r} \wedge \mathbf{s})]\mathbf{q}.$$

(c) yields $(\mathbf{p} \wedge \mathbf{q}).(\mathbf{r} \wedge \mathbf{s}) = \mathbf{p}.[\mathbf{q} \wedge (\mathbf{r} \wedge \mathbf{s})]$ by considering $\mathbf{r} \wedge \mathbf{s}$ as a single vector and interchanging the dot and cross products in the resultant scalar triple product. In this way it reduces to case (a).

(d) Consider $\mathbf{t} \equiv (\mathbf{p} \wedge \mathbf{q}) \wedge (\mathbf{r} \wedge \mathbf{s})$. Since $\mathbf{p} \wedge \mathbf{q}$ is perpendicular to the plane of $\mathbf{p}$ and $\mathbf{q}$, it follows that $\mathbf{t}$ lies in the plane of $\mathbf{p}$ and $\mathbf{q}$. Similarly, it lies in the plane of $\mathbf{r}$ and $\mathbf{s}$. Hence it must be parallel to the intersection of the plane containing $\mathbf{p}$ and $\mathbf{q}$ with the plane containing $\mathbf{r}$ and $\mathbf{s}$. Taking now $(\mathbf{p} \wedge \mathbf{q})$ as a single vector, $\mathbf{t}$ may be expanded to yield

$$\mathbf{t} = (\mathbf{p} \wedge \mathbf{q}) \wedge (\mathbf{r} \wedge \mathbf{s}) = [(\mathbf{p} \wedge \mathbf{q}).\mathbf{s}]\mathbf{r} - [(\mathbf{p} \wedge \mathbf{q}).\mathbf{r}]\mathbf{s}. \quad (50)$$

Alternatively, taking $\mathbf{r} \wedge \mathbf{s}$ as a single vector,

$$\mathbf{t} = (\mathbf{r} \wedge \mathbf{s}) \wedge (\mathbf{q} \wedge \mathbf{p}) = [(\mathbf{r} \wedge \mathbf{s}).\mathbf{p}]\mathbf{q} - [(\mathbf{r} \wedge \mathbf{s}).\mathbf{q}]\mathbf{p}. \quad (51)$$

Equating the alternative forms for $\mathbf{t}$ given by eqns. (50) and (51) yields

$$\mathbf{r} = \frac{[\mathbf{r}.(\mathbf{q} \wedge \mathbf{s})]\mathbf{p}}{[\mathbf{p}.(\mathbf{q} \wedge \mathbf{s})]} + \frac{[\mathbf{r}.(\mathbf{s} \wedge \mathbf{p})]\mathbf{q}}{[\mathbf{p}.(\mathbf{q} \wedge \mathbf{s})]} + \frac{[\mathbf{r}.(\mathbf{p} \wedge \mathbf{q})]\mathbf{s}}{[\mathbf{p}.(\mathbf{q} \wedge \mathbf{s})]}. \quad (52)$$

Let us now compare eqn. (52) with eqn. (7) in which we considered the resolution of a vector $\mathbf{F}$ into components parallel to $\mathbf{p}$, $\mathbf{q}$ and $\mathbf{s}$. It is readily seen that these two equations are identical if $\mathbf{F} = \mathbf{r}$ and

$$a = \frac{[\mathbf{r}.(\mathbf{q} \wedge \mathbf{s})]}{[\mathbf{p}.(\mathbf{q} \wedge \mathbf{s})]}, \ b = \frac{[\mathbf{r}.(\mathbf{s} \wedge \mathbf{p})]}{[\mathbf{p}.(\mathbf{q} \wedge \mathbf{s})]}, \ c = \frac{[\mathbf{r}.(\mathbf{p} \wedge \mathbf{q})]}{[\mathbf{p}.(\mathbf{q} \wedge \mathbf{s})]}. \quad (53)$$

In other words, these are the explicit expressions for a, b and c in terms of $\mathbf{p}$, $\mathbf{q}$, $\mathbf{r}$ and $\mathbf{s}$ which follows from eqn. (7).

Worked examples

(1) Prove $(\mathbf{p} \wedge \mathbf{q}).[(\mathbf{q} \wedge \mathbf{r}) \wedge (\mathbf{r} \wedge \mathbf{p})] = [\mathbf{p}.(\mathbf{q} \wedge \mathbf{r})]^2$.

Ans. From eqn. (50), $(\mathbf{q} \wedge \mathbf{r}) \wedge (\mathbf{r} \wedge \mathbf{p}) = [(\mathbf{q} \wedge \mathbf{r}).\mathbf{p}]\mathbf{r}$, since $(\mathbf{q} \wedge \mathbf{r}).\mathbf{r} = 0$.

D

Therefore

$$(\mathbf{p} \wedge \mathbf{q})[(\mathbf{q} \wedge \mathbf{r}) \wedge (\mathbf{r} \wedge \mathbf{p})] = [(\mathbf{q} \wedge \mathbf{r}).\mathbf{p}][(\mathbf{p} \wedge \mathbf{q}).\mathbf{r}] = [\mathbf{p}.(\mathbf{q} \wedge \mathbf{r})]^2.$$

(2) Prove that:

$$[\mathbf{p}.(\mathbf{q} \wedge \mathbf{r})][\mathbf{s}.(\mathbf{t} \wedge \mathbf{u})] = \begin{vmatrix} \mathbf{p}.\mathbf{s} & \mathbf{p}.\mathbf{t} & \mathbf{p}.\mathbf{u} \\ \mathbf{q}.\mathbf{s} & \mathbf{q}.\mathbf{t} & \mathbf{q}.\mathbf{u} \\ \mathbf{r}.\mathbf{s} & \mathbf{r}.\mathbf{t} & \mathbf{r}.\mathbf{u} \end{vmatrix}.$$

Ans. From eqn. (42a)

$$\mathbf{p}.(\mathbf{q} \wedge \mathbf{r}) = \begin{vmatrix} p_x & p_y & p_z \\ q_x & q_y & q_z \\ r_x & r_y & r_z \end{vmatrix}$$

and

$$\mathbf{s}.(\mathbf{t} \wedge \mathbf{u}) = \begin{vmatrix} s_x & t_x & u_x \\ s_y & t_y & u_y \\ s_z & t_z & u_z \end{vmatrix}$$

since a determinant is unaltered by interchange of rows and columns. Therefore by the rule for the product of two determinants,

$$[\mathbf{p}.(\mathbf{q} \wedge \mathbf{r})][\mathbf{s}.(\mathbf{t} \wedge \mathbf{u})] = \begin{vmatrix} p_x & p_y & p_z \\ q_x & q_y & q_z \\ r_x & r_y & r_z \end{vmatrix} \times \begin{vmatrix} s_x & t_x & u_x \\ s_y & t_y & u_y \\ s_z & t_z & u_z \end{vmatrix}$$

$$= \begin{vmatrix} p_x s_x + p_y s_y + p_z s_z & p_x t_x + p_y t_y + p_z t_z & p_x u_x + p_y u_y + p_z u_z \\ q_x s_x + q_y s_y + q_z s_z & q_x t_x + q_y t_y + q_z t_z & q_x u_x + q_y u_y + q_z u_z \\ r_x s_x + r_y s_y + r_z s_z & r_x t_x + r_y t_y + r_z t_z & r_x u_x + r_y u_y + r_z u_z \end{vmatrix}$$

$$= \begin{vmatrix} \mathbf{p}.\mathbf{s} & \mathbf{p}.\mathbf{t} & \mathbf{p}.\mathbf{u} \\ \mathbf{q}.\mathbf{s} & \mathbf{q}.\mathbf{t} & \mathbf{q}.\mathbf{u} \\ \mathbf{r}.\mathbf{s} & \mathbf{r}.\mathbf{t} & \mathbf{r}.\mathbf{u} \end{vmatrix}.$$

Exercises

(1) Find the value of a such that $2\mathbf{i} + a\mathbf{j} - 4\mathbf{k}, \mathbf{i} - 3a\mathbf{j} + 2\mathbf{k}, -\mathbf{i} + \mathbf{j} + \mathbf{k}$ are coplanar vectors.

(2) By taking the triple scalar product of both sides of eqn. (7) with vectors $\mathbf{p}$ and $\mathbf{q}$, $\mathbf{q}$ and $\mathbf{s}$, $\mathbf{s}$ and $\mathbf{p}$ in turn, obtain the values of a, b, c given in eqn. (53).

(3) Corresponding to vectors $\mathbf{p}$, $\mathbf{q}$, $\mathbf{r}$, the vectors $\mathbf{p}'$, $\mathbf{q}'$, $\mathbf{r}'$ are defined by

$$\mathbf{p}' = \frac{\mathbf{q} \wedge \mathbf{r}}{\mathbf{p}.(\mathbf{q} \wedge \mathbf{r})}, \quad \mathbf{q}' = \frac{\mathbf{r} \wedge \mathbf{p}}{\mathbf{p}.(\mathbf{q} \wedge \mathbf{r})}, \quad \mathbf{r}' = \frac{\mathbf{p} \wedge \mathbf{q}}{\mathbf{p}.(\mathbf{q} \wedge \mathbf{r})}.$$

Prove that

(a) $\mathbf{p}.\mathbf{p}' = \mathbf{q}.\mathbf{q}' = \mathbf{r}.\mathbf{r}' = 1$ and

$\mathbf{p}.\mathbf{q}' = \mathbf{p}.\mathbf{r}' = \mathbf{q}.\mathbf{p}' = \mathbf{q}.\mathbf{r}' = \mathbf{r}.\mathbf{p}' = \mathbf{r}.\mathbf{q}' = 0.$

(b) $[\mathbf{p}.(\mathbf{q} \wedge \mathbf{r})][\mathbf{p}'.(\mathbf{q}' \wedge \mathbf{r}')] = 1.$

(c) $\mathbf{p} = \dfrac{\mathbf{q}' \wedge \mathbf{r}'}{\mathbf{p}'.(\mathbf{q}' \wedge \mathbf{r}')}, \quad \mathbf{q} = \dfrac{\mathbf{r}' \wedge \mathbf{p}'}{\mathbf{p}'.(\mathbf{q}' \wedge \mathbf{r}')}, \quad \mathbf{r} = \dfrac{\mathbf{p}' \wedge \mathbf{q}'}{\mathbf{p}.(\mathbf{q}' \wedge \mathbf{r}')}.$

(4) Prove that $(\mathbf{p} \wedge \mathbf{q}).[(\mathbf{p} \wedge \mathbf{r}) \wedge \mathbf{s}] = (\mathbf{p}.\mathbf{s})[\mathbf{p}.(\mathbf{q} \wedge \mathbf{r})].$

4

Differentiation of Vectors

4.1 THE DERIVATIVE OF A VECTOR

In all our previous work, we have been concerned with vectors which were constant and did not vary. However, in many physical contexts one is interested in vectors which alter with position or time: for example, the velocity **v** of a particle may change with time t, or the strength **E** of an electric field may depend on position x. Such a situation is the vector analogue of a scalar dependent variable y being a function of another scalar variable x, which is usually represented by $y = y(x)$. Similarly, if a vector **F** depends on a scalar variable u, this relationship may be denoted by $\mathbf{F} = \mathbf{F}(u)$, and it is the concept of differentiation with respect to u which we will be concerned with in this chapter, before proceeding in the next chapter to the case where the independent variable is itself a vector.

The definition of differentiation with respect to the independent scalar variable may be explained by reference to Fig. 28 in which $\overrightarrow{OP} = \mathbf{F}(u)$ and $\overrightarrow{OQ} = \mathbf{F}(u + \delta u)$, where δu is a small increment in u. $\overrightarrow{PQ} = \delta \mathbf{F}(u) = \mathbf{F}(u + \delta u) - \mathbf{F}(u)$ from the definition of vector subtraction, and as $\delta u \to 0$,

$$\frac{\delta \mathbf{F}}{\delta u} = \frac{\mathbf{F}(u + \delta u) - \mathbf{F}(u)}{\delta u}$$

tends to a limiting value denoted by dF/du, and termed " the derivative of F with respect to u "; i.e.

$$\frac{dF}{du} = \lim_{\delta u \to 0} \frac{F(u + \delta u) - F(u)}{\delta u}. \tag{54}$$

Since PQ is a chord of the curve described by the terminal point P of F, it is clear that as $\delta u \to 0$ the direction of PQ

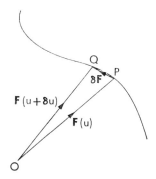

FIG. 28.

will tend to the tangent at P, and hence the direction of dF/du is along the tangent to the locus of P.

Since dF/du is itself a vector, it is clear that definition (54) may be applied repeatedly to define higher order differential coefficients as long as the corresponding limits exist. Thus, for example,

$$\frac{d^2F}{du^2} = \lim_{\delta u \to 0} \frac{[\partial F/\partial u](u + \delta u) - [\partial F/\partial u](u)}{\delta u}.$$

4.2 DIFFERENTIATION OF SUMS AND PRODUCTS

We list the following results before proving them. If **F**, **G** and **H** are vector functions and y is a scalar function of a scalar independent variable u,

$$\frac{d}{du}(\mathbf{F} + \mathbf{G}) = \frac{d\mathbf{F}}{du} + \frac{d\mathbf{G}}{du}, \tag{55}$$

$$\frac{d}{du}(\mathbf{F}.\mathbf{G}) = \mathbf{F}.\frac{d\mathbf{G}}{du} + \frac{d\mathbf{F}}{du}.\mathbf{G}, \tag{56}$$

$$\frac{d}{du}(\mathbf{F} \wedge \mathbf{G}) = \mathbf{F} \wedge \frac{d\mathbf{G}}{du} + \frac{d\mathbf{F}}{du} \wedge \mathbf{G}, \tag{57}$$

$$\frac{d}{du}(y\mathbf{F}) = y\frac{d\mathbf{F}}{du} + \frac{dy}{du}\mathbf{F}, \tag{58}$$

$$\frac{d}{du}[\mathbf{F}.(\mathbf{G} \wedge \mathbf{H})] = \mathbf{F}.\left(\mathbf{G} \wedge \frac{d\mathbf{H}}{du}\right) + \mathbf{F}.\left(\frac{d\mathbf{G}}{du} \wedge \mathbf{H}\right) + \\ + \frac{d\mathbf{F}}{du}.(\mathbf{G} \wedge \mathbf{H}), \tag{59}$$

$$\frac{d}{du}[\mathbf{F} \wedge (\mathbf{G} \wedge \mathbf{H})] = \mathbf{F} \wedge \left(\mathbf{G} \wedge \frac{d\mathbf{H}}{du}\right) + \\ + \mathbf{F} \wedge \left(\frac{d\mathbf{G}}{du} \wedge \mathbf{H}\right) + \frac{d\mathbf{F}}{du} \wedge (\mathbf{G} \wedge \mathbf{H}), \tag{60}$$

$$\frac{d\mathbf{F}}{ds} = \frac{d\mathbf{F}}{du} \times \frac{du}{ds}, \tag{61}$$

where in (61) u is itself a function of another scalar s. It should be noted that in the results involving vector products, the order of the factors is, of course, important. The proofs of these results are similar to the corresponding proofs in elementary calculus.

Proofs.

$$(55) \quad \frac{d}{du}(\mathbf{F} + \mathbf{G}) = \lim_{\delta u \to 0} \frac{\delta(\mathbf{F} + \mathbf{G})}{\delta u} = \lim_{\delta u \to 0} \left(\frac{\delta \mathbf{F}}{\delta u} + \frac{\delta \mathbf{G}}{\delta u}\right)$$

$$= \frac{d\mathbf{F}}{du} + \frac{d\mathbf{G}}{du}.$$

$$(56) \quad \frac{d}{du}(\mathbf{F}.\mathbf{G}) = \lim_{\delta u \to 0} \frac{(\mathbf{F} + \delta \mathbf{F}).(\mathbf{G} + \delta \mathbf{G}) - \mathbf{F}.\mathbf{G}}{\delta u}$$

$$= \lim_{\delta u \to 0} \frac{\delta \mathbf{F}.\mathbf{G} + \mathbf{F}.\delta \mathbf{G} + \delta \mathbf{F}.\delta \mathbf{G}}{\delta u}$$

$$= \frac{d\mathbf{F}}{du}.\mathbf{G} + \mathbf{F}.\frac{d\mathbf{G}}{du}.$$

(57) is identical with (56) if the scalar product is replaced by a vector product.

(58) is identical with (56) if the scalar product is replaced by an ordinary product of scalar and vector.

$$(59) \quad \frac{d}{du}[\mathbf{F}.(\mathbf{G} \wedge \mathbf{H})]$$

$$= \lim_{\delta u \to 0} \frac{(\mathbf{F} + \delta \mathbf{F}).[(\mathbf{G} + \delta \mathbf{G}) \wedge (\mathbf{H} + \delta \mathbf{H})] - \mathbf{F}.(\mathbf{G} \wedge \mathbf{H})}{\delta u}$$

$$= \lim_{\delta u \to 0} \frac{\mathbf{F}.(\mathbf{G} \wedge \delta \mathbf{H}) + \mathbf{F}.(\delta \mathbf{G} \wedge \mathbf{H}) + \delta \mathbf{F}.(\mathbf{G} \wedge \mathbf{H})}{\delta u}$$

$$\frac{+ \text{ terms involving products of } \delta\text{'s}}{\delta u}$$

$$= \mathbf{F}.\left(\mathbf{G} \wedge \frac{d\mathbf{H}}{du}\right) + \mathbf{F}.\left(\frac{d\mathbf{G}}{du} \wedge \mathbf{H}\right) + \frac{d\mathbf{F}}{du}.(\mathbf{G} \wedge \mathbf{H}).$$

(60) is identical with (59) if the first scalar product is replaced by a vector product.

$$(61) \quad \frac{d\mathbf{F}}{ds} = \lim_{\delta s \to 0} \frac{\delta \mathbf{F}}{\delta s} = \lim_{\delta s \to 0} \frac{\delta \mathbf{F}}{\delta u} \times \frac{\delta u}{\delta s} = \frac{d\mathbf{F}}{du} \times \frac{du}{ds}.$$

Worked examples

(1) Show that $\mathbf{F} \cdot \dfrac{d\mathbf{F}}{du} = F \dfrac{dF}{du}$ and deduce that if $\mathbf{F}$ has constant magnitude, $\mathbf{F}$ and $d\mathbf{F}/du$ are perpendicular. Prove the latter result geometrically.

Ans. Since

$$\mathbf{F}.\mathbf{F} = F^2, \ d(\mathbf{F}.\mathbf{F})/du = dF^2/du. \tag{62}$$

Further, from result (56) above,

$$d(\mathbf{F}.\mathbf{F})/du = \mathbf{F}.(d\mathbf{F}/du) + (d\mathbf{F}/du).\mathbf{F} = 2\mathbf{F}.(d\mathbf{F}/du).$$

Therefore

$$2\mathbf{F} \cdot \frac{d\mathbf{F}}{du} = \frac{dF^2}{du} = 2F\frac{dF}{du}, \quad \text{and so} \quad \mathbf{F} \cdot \frac{d\mathbf{F}}{du} = F\frac{dF}{du}. \tag{63}$$

If F is constant, $dF/du = 0$ and therefore it follows from the above result (63) that $\mathbf{F}.(d\mathbf{F}/du) = 0$. Hence since $\mathbf{F}$ and $d\mathbf{F}/du$ are both non-zero, they must be perpendicular.

Alternatively, if $\overrightarrow{OP} = \mathbf{F}$, the locus of P is a circle as in Fig. 29, since F is constant. $d\mathbf{F}/du$ is along the tangent PT at P, and since PT is perpendicular to OP the result follows.

(2) Simplify $\dfrac{d}{du}\left[\mathbf{F} \cdot \left(\dfrac{d\mathbf{F}}{du} \wedge \dfrac{d^2\mathbf{F}}{du^2} \right) \right].$

Ans. Applying formula (59) of this section

$$\frac{d}{du}\left[\mathbf{F} \cdot \left(\frac{d\mathbf{F}}{du} \wedge \frac{d^2\mathbf{F}}{du^2} \right) \right] = \mathbf{F} \cdot \left(\frac{d\mathbf{F}}{du} \wedge \frac{d^3\mathbf{F}}{du^3} \right) + \mathbf{F} \cdot \left(\frac{d^2\mathbf{F}}{du^2} \wedge \frac{d^2\mathbf{F}}{du^2} \right) +$$
$$+ \frac{d\mathbf{F}}{du} \cdot \left(\frac{d\mathbf{F}}{du} \wedge \frac{d^2\mathbf{F}}{du^2} \right).$$

Now, the final two terms of this expression are zero since each is a triple scalar product with two identical factors. Thus,

$$\frac{\mathrm{d}}{\mathrm{d}u}\left[\mathbf{F} \cdot \left(\frac{\mathrm{d}\mathbf{F}}{\mathrm{d}u}\wedge\frac{\mathrm{d}^2\mathbf{F}}{\mathrm{d}u^2}\right) \right] = \mathbf{F} \cdot \left(\frac{\mathrm{d}\mathbf{F}}{\mathrm{d}u}\wedge\frac{\mathrm{d}^3\mathbf{F}}{\mathrm{d}u^3}\right).$$

Exercise

Simplify:

(1) $\dfrac{\mathrm{d}}{\mathrm{d}u}\left(\mathbf{F} \wedge \dfrac{\mathrm{d}\mathbf{F}}{\mathrm{d}u}\right).$

(2) $\dfrac{\mathrm{d}}{\mathrm{d}u}\left(\mathbf{F} \wedge \dfrac{\mathrm{d}\mathbf{G}}{\mathrm{d}u} + \mathbf{G} \wedge \dfrac{\mathrm{d}\mathbf{F}}{\mathrm{d}u}\right).$

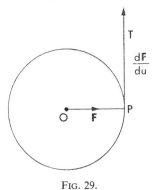

Fig. 29.

4.3 COMPONENTS OF A DERIVATIVE

If $\mathbf{F}$ is written in terms of its cartesian components, i.e. $\mathbf{F} = F_x\mathbf{i} + F_y\mathbf{j} + F_z\mathbf{k}$, what is $\mathrm{d}\mathbf{F}/\mathrm{d}u$ in terms of the derivatives of F_x, F_y, F_z? Now, in its cartesian form $\mathbf{F}$ is written as the sum of three vectors, i.e. $F_x\mathbf{i}$, $F_y\mathbf{j}$, $F_z\mathbf{k}$, each of which is the product of a variable scalar and a constant vector. Therefore, applying formulae (55) and (58) of the last section, it follows immediately that

$$\frac{\mathrm{d}\mathbf{F}}{\mathrm{d}u} = \frac{\mathrm{d}F_x}{\mathrm{d}u}\mathbf{i} + \frac{\mathrm{d}F_y}{\mathrm{d}u}\mathbf{j} + \frac{\mathrm{d}F_z}{\mathrm{d}u}\mathbf{k}. \tag{64}$$

This form for $d\mathbf{F}/du$ may be used to prove any of the formulae given in the last section, since it effectively reduces the vector equalities to scalar equalities between the resolutes in each direction. Thus, for example,

$$\frac{d}{du}(\mathbf{F} \wedge \mathbf{G}) = \mathbf{i}\frac{d(F_y G_z - F_z G_y)}{du} + \mathbf{j}\frac{d(F_z G_x - F_x G_z)}{du} +$$

$$+ \mathbf{k}\frac{d(F_x G_y - F_y G_x)}{du}$$

$$= \mathbf{i}\left(F_y\frac{dG_z}{du} + G_z\frac{dF_y}{du} - F_z\frac{dG_y}{du} - G_y\frac{dF_z}{du}\right) +$$

$$+ \text{ similar terms in } \mathbf{j}, \mathbf{k}.$$

$$= \left[\mathbf{i}\left(F_y\frac{dG_z}{du} - F_z\frac{dG_y}{du}\right) + \text{ similar terms in } \mathbf{j}, \mathbf{k}\right] +$$

$$+ \left[\mathbf{i}\left(G_z\frac{dF_y}{du} - G_y\frac{dF_z}{du}\right) + \text{ similar terms in } \mathbf{j}, \mathbf{k}\right]$$

$$= \mathbf{F} \wedge \frac{d\mathbf{G}}{du} + \frac{d\mathbf{F}}{du} \wedge \mathbf{G}.$$

Worked examples

If $\mathbf{F} = e^{3u}\mathbf{i} + u^2\mathbf{j} - \ln(1 + u)\mathbf{k}$, find

(a) $d\mathbf{F}/du$,   (c) $|d^2\mathbf{F}/du^2|$,

(b) $d^2\mathbf{F}/du^2$,   (d) $d(\mathbf{F}.\mathbf{F})/du$, at $u = 0$.

Ans. (a) $d\mathbf{F}/du = 3e^{3u}\mathbf{i} + 2u\mathbf{j} - (1 + u)^{-1}\mathbf{k}$.

(b) $d^2\mathbf{F}/du^2 = 9e^{3u}\mathbf{i} + 2\mathbf{j} + (1 + u)^{-2}\mathbf{k}$.

Therefore at $u = 0$, $d\mathbf{F}/du = 3\mathbf{i} - \mathbf{k}$, $d^2\mathbf{F}/du^2 = 9\mathbf{i} + 2\mathbf{j} + \mathbf{k}$ and (c) $|d^2\mathbf{F}/du^2| = (9^2 + 2^2 + 1)^{\frac{1}{2}} = \sqrt{86}$.

(d) $\mathbf{F}.\mathbf{F} = e^{6u} + u^4 + [\ln(1 + u)]^2$.

Therefore

$$d(\mathbf{F}.\mathbf{F})/du = 6e^{6u} + 4u^3 + 2\ln(1 + u)/(1 + u)$$

and at $u = 0$, $d(\mathbf{F}.\mathbf{F})/du = 6$.

Exercises

(1) Prove formula (59) of the last section by expressing each vector in its cartesian form.

(2) If $\mathbf{F} = u\mathbf{i} - (1 + u^2)\mathbf{j} + 2u\mathbf{k}$ and $G = 4\mathbf{i} + (3u + 1)\mathbf{j} + u^2\mathbf{k}$, calculate

(a) $\dfrac{d\mathbf{F}.\mathbf{G}}{du}$,   (b) $\dfrac{d\mathbf{F} \wedge \mathbf{G}}{du}$,   (c) $\dfrac{d(\mathbf{F} + \mathbf{G})}{du}$,

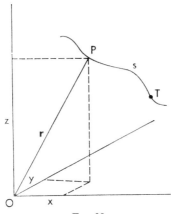

FIG. 30.

(d) $\dfrac{d|\mathbf{F} + \mathbf{G}|}{du}$,   (e) $\dfrac{d[\mathbf{F}.(d\mathbf{G}/du)]}{du}$

at $u = 1$.

[*Remark*: These can be obtained from $d\mathbf{F}/du$ and $d\mathbf{G}/du$ using the formulae of the last section; alternatively, the quantity to be differentiated can be evaluated explicitly before differentiation.]

Application to Differential Geometry

If $\mathbf{r} = \overrightarrow{OP}$ where O is a fixed origin and P is a variable point, then $r_x = x$, $r_y = y$, $r_z = z$, where x, y and z are the usual three-dimensional cartesian coordinates as shown in Fig. 30. If $\mathbf{r} = \mathbf{r}(u)$, P will describe some general curve in three dimensions as u varies, for which the vector equation may be written

$$\mathbf{i}x + \mathbf{j}y + \mathbf{k}z = \mathbf{i}x(u) + \mathbf{j}y(u) + \mathbf{k}z(u),$$

which is equivalent to $x = x(u), y = y(u), z = z(u)$: this in turn is a parametric form for the cartesian equation of the curve. As explained earlier, $d\mathbf{r}/du$ at any point will be parallel to the tangent at that point, i.e. the tangent will be parallel to the vector $\mathbf{i}(dx/du) + \mathbf{j}(dy/du) + \mathbf{k}(dz/du)$. Now, suppose that $u = s$, the arc length of the curve measured from a fixed point T on it as shown in Fig. 30. Then $d\mathbf{r}/ds$ at P is parallel to the tangent at P, and also

$$\left| d\mathbf{r}/ds \right| = [(dx/ds)^2 + (dy/ds)^2 + (dz/ds)^2]^{\frac{1}{2}} \quad (65)$$

since $d\mathbf{r}/ds = \mathbf{i}(dx/ds) + \mathbf{j}(dy/ds) + \mathbf{k}(dz/ds)$. Now, for an element δs of the curve, $(\delta s)^2 = (\delta x)^2 + (\delta y)^2 + (\delta z)^2$, and thus

$$(dx/ds)^2 + (dy/ds)^2 + (dz/ds)^2 = 1$$

so that eqn. (65) yields $\left| d\mathbf{r}/ds \right| = 1$. Hence $d\mathbf{r}/ds$ at any point is a unit vector along the tangent at that point. If we denote this unit tangent vector by $\mathbf{T}$, it follows from example (1) of the last section that since $\mathbf{T}.\mathbf{T} = 1$, $\mathbf{T}.(d\mathbf{T}/ds) = 0$, and therefore $d\mathbf{T}/ds(= d^2\mathbf{r}/ds^2)$ is perpendicular to $\mathbf{T}$. If $\mathbf{N}$ is a unit vector parallel to $d\mathbf{T}/ds$ (and therefore perpendicular to $\mathbf{T}$) we may write $d\mathbf{T}/ds = \mathbf{N}/\rho$ where the scalar quantity ρ is called the radius of curvature; $\mathbf{N}$ is called the principal normal. Clearly $\rho = \left| d\mathbf{T}/ds \right|^{-1}$.

Worked example

Find the unit tangent vector at the point $(3, 0, 3)$ for the curve with parametric cartesian equations

$$x = 3u, \quad y = u^2 - u, \quad z = 2u^2 + 1.$$

Ans. If $\mathbf{r} = \mathbf{i}x + \mathbf{j}y + \mathbf{k}z$, the unit tangent vector is

$$\frac{(\mathrm{d}\mathbf{r}/\mathrm{d}u)}{\left|\mathrm{d}\mathbf{r}/\mathrm{d}u\right|}.$$

Now

$$\mathbf{r} = 3u\mathbf{i} + (u^2 - u)\mathbf{j} + (2u^2 + 1)\mathbf{k}$$

and hence

$$\mathrm{d}\mathbf{r}/\mathrm{d}u = 3\mathbf{i} + (2u - 1)\mathbf{j} + 4u\mathbf{k}.$$

The point $(3, 0, 3)$ corresponds to $u = 1$, and here

$$\mathrm{d}\mathbf{r}/\mathrm{d}u = 3\mathbf{i} + \mathbf{j} + 4\mathbf{k}$$

and

$$\left|\mathrm{d}\mathbf{r}/\mathrm{d}u\right| = (3^2 + 1 + 4^2)^{\frac{1}{2}} = \sqrt{26}.$$

Thus the required unit tangent vector is $\mathbf{T} = 26^{-\frac{1}{2}}(3\mathbf{i} + \mathbf{j} + 4\mathbf{k})$.

Exercise

Find the unit tangent vector and the radius of curvature at any point for the curve with parametric cartesian equations

$$x = 5\cos u, \quad y = 5\sin u, \quad z = 12u.$$

$\Bigg[$ *Hint.* From formula (61) of the preceeding section,

$$\frac{\mathrm{d}\mathbf{T}}{\mathrm{d}s} = \frac{\mathrm{d}\mathbf{T}}{\mathrm{d}u} \times \frac{\mathrm{d}u}{\mathrm{d}s}, \quad \text{and} \quad \frac{\mathrm{d}s}{\mathrm{d}u} = \left|\frac{\mathrm{d}\mathbf{r}}{\mathrm{d}u}\right|. \Bigg]$$

4.4 APPLICATIONS TO MECHANICS

Velocity and Acceleration

Consider a point P moving along a curve in three dimensions, and let $\overrightarrow{OP} = \mathbf{r}$ where O is a convenient origin. Clearly $\mathbf{r}$ is a function of the time t and the velocity $\mathbf{v}$ of P may be defined by $\mathbf{v} = \mathrm{d}\mathbf{r}/\mathrm{d}t$; the acceleration $\mathbf{a}$ is defined by $\mathbf{a} = \mathrm{d}\mathbf{v}/\mathrm{d}t =$

$d^2\mathbf{r}/dt^2$. As shown earlier $\mathbf{v}$ will be along the tangent to the curve, and if $\mathbf{r} = \mathbf{i}x + \mathbf{j}y + \mathbf{k}z$,

$$\mathbf{v} = \mathbf{i}\frac{dx}{dt} + \mathbf{j}\frac{dy}{dt} + \mathbf{k}\frac{dz}{dt} \quad \text{and} \quad \mathbf{a} = \mathbf{i}\frac{d^2x}{dt^2} + \mathbf{j}\frac{d^2y}{dt^2} + \mathbf{k}\frac{d^2z}{dt^2}.$$

Worked example

A particle moves so that after time t, $x = 3t - 4$, $y = 2t^2 + t$, $z = 3t^3 - 6t$. Find the resolutes of its velocity and acceleration in the direction of $3\mathbf{i} + \mathbf{j} - 2\mathbf{k}$ for $t = 2$.

Ans. $\mathbf{r} = \mathbf{i}(3t - 4) + \mathbf{j}(2t^2 + t) + \mathbf{k}(3t^3 - 6t)$,

$d\mathbf{r}/dt = 3\mathbf{i} + (4t + 1)\mathbf{j} + \mathbf{k}(9t^2 - 6)$,

$d^2\mathbf{r}/dt^2 = 4\mathbf{j} + 18t\mathbf{k}$.

Therefore at $t = 2$,

$$d\mathbf{r}/dt = 3\mathbf{i} + 9\mathbf{j} + 30\mathbf{k}, \quad d^2\mathbf{r}/dt^2 = 4\mathbf{j} + 36\mathbf{k}.$$

Now a unit vector $\mathbf{l}$ parallel to $3\mathbf{i} + \mathbf{j} - 2\mathbf{k}$ is $(3\mathbf{i} + \mathbf{j} - 2\mathbf{k})/|3\mathbf{i} + \mathbf{j} - 2\mathbf{k}| = 14^{-\frac{1}{2}}(3\mathbf{i} + \mathbf{j} - 2\mathbf{k})$, and the resolute of a vector $\mathbf{F}$ in the direction of $\mathbf{l} = F\cos\theta = \mathbf{F}.\mathbf{l}$, where θ is the angle between $\mathbf{F}$ and $\mathbf{l}$. Thus the required resolute of velocity

$$= 14^{-\frac{1}{2}}(3\mathbf{i} + \mathbf{j} - 2\mathbf{k}).(3\mathbf{i} + 9\mathbf{j} + 30\mathbf{k})$$

$$= 14^{-\frac{1}{2}}(3 \times 3 + 1 \times 9 - 2 \times 30) = -42/\sqrt{14}.$$

The required resolute of acceleration equals

$$14^{-\frac{1}{2}}(3\mathbf{i} + \mathbf{j} - 2\mathbf{k}).(4\mathbf{j} + 36\mathbf{k})$$

$$= 14^{-\frac{1}{2}}(1 \times 4 - 2 \times 36) = -68/\sqrt{14}.$$

Exercise

Find the magnitude of the velocity and acceleration after time t of a point moving along a curve such that

$$x = 3\cos t - \sin t, \quad y = 3\sin t + 4t, \quad z = 2t^2.$$

Normal and Tangential Components of Acceleration

We now proceed to show that the acceleration of a point moving along a general curve may be resolved into a component dv/dt along the tangent and a component v^2/ρ along the principal normal; that is

$$\mathbf{a} = (dv/dt)\mathbf{T} + (v^2/\rho)\mathbf{N}.$$

Letting $\mathbf{v} = v\mathbf{T}$,

$$\frac{d\mathbf{v}}{dt} = v\frac{d\mathbf{T}}{dt} + \frac{dv}{dt}\mathbf{T} \tag{66}$$

making use of formula (58) of §4.2. Also

$$\frac{d\mathbf{T}}{dt} = \frac{d\mathbf{T}}{ds} \times \frac{ds}{dt} = \frac{\mathbf{N}}{\rho} \times v$$

since

$$v = [(dx/dt)^2 + (dy/dt)^2 + (dz/dt)^2]^{\frac{1}{2}} = ds/dt.$$

Hence, substituting into eqn. (66),

$$\frac{d\mathbf{v}}{dt} = \mathbf{a} = \frac{v^2}{\rho}\mathbf{N} + \frac{dv}{dt}\mathbf{T}. \tag{67}$$

Radial and Transverse Components of Acceleration

For motion along a curve in two dimensions described by polar coordinates (r, θ) it is of interest to obtain the radial (parallel to $\mathbf{r}$) and transverse (perpendicular to $\mathbf{r}$) resolutes of the acceleration. Now, since motion is in two dimensions

$$\mathbf{r} = \mathbf{i}x + \mathbf{j}y = \mathbf{i}r\cos\theta + \mathbf{j}r\sin\theta,$$

and therefore

$$\dot{\mathbf{r}} = \mathbf{i}(\dot{r}\cos\theta - r\sin\theta\dot{\theta}) + \mathbf{j}(\dot{r}\sin\theta + r\cos\theta\dot{\theta}),$$

where differentiation with respect to t is denoted by a dot over the letter.

Therefore

$$\ddot{\mathbf{r}} = \mathbf{i}(\ddot{r}\cos\theta - \dot{r}\sin\theta\dot{\theta} - \dot{r}\sin\theta\dot{\theta} - r\cos\theta\dot{\theta}\dot{\theta} - r\sin\theta\ddot{\theta}) +$$
$$+ \mathbf{j}(\ddot{r}\sin\theta + \dot{r}\cos\theta\dot{\theta} + \dot{r}\cos\theta\dot{\theta} - r\sin\theta\dot{\theta}\dot{\theta} + r\cos\theta\ddot{\theta})$$
$$= \mathbf{i}[(\ddot{r} - r\dot{\theta}^2)\cos\theta - (2\dot{r}\dot{\theta} + r\ddot{\theta})\sin\theta] +$$
$$+ \mathbf{j}[(\ddot{r} - r\dot{\theta}^2)\sin\theta + (2\dot{r}\dot{\theta} + r\ddot{\theta})\cos\theta]. \tag{68}$$

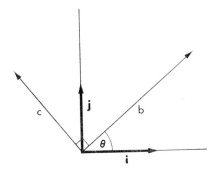

Fig. 31.

Referring to Fig. 31, it is clear that if b and c are the radial and transverse resolutes of acceleration, then

$$\ddot{\mathbf{r}} = \mathbf{i}(b\cos\theta - c\sin\theta) + \mathbf{j}(b\sin\theta + c\cos\theta). \tag{69}$$

Comparing eqns. (68) and (69) it is seen that

$$b = \ddot{r} - r\dot{\theta}^2, \quad c = r\ddot{\theta} + 2\dot{r}\dot{\theta} = r^{-1}\mathrm{d}(r^2\dot{\theta})/\mathrm{d}t. \tag{70}$$

Dynamics of a System of Particles

We can now prove two basic results concerning the dynamics of a system of particles:

(A) The total external force acting on a system of particles is equal to the total mass times the acceleration of the c.g.

(*B*) The total moment of the external forces about any point is equal to the rate of change of the angular momentum of the system about that point.

Proofs. (*A*) Suppose the system to consist of N particles each of mass $m_p (1 \leqq p \leqq N)$ and at position vector $\mathbf{r}_p$, with respect to an origin O. Then for a single particle of the system

$$\mathbf{F}_p + \mathbf{F}_p{}' = d(m_p \dot{\mathbf{r}}_p)/dt = m_p \ddot{\mathbf{r}}_p \tag{71}$$

from Newton's Second Law, where $\mathbf{F}_p$ is the external force acting on m_p and $\mathbf{F}_p{}'$ is the corresponding internal force, i.e. the force acting on m_p due to its interaction with neighbouring particles. Summing equation (71) for all particles in the system yields

$$\sum_{p=1}^{N} \mathbf{F}_p + \sum_{p=1}^{N} \mathbf{F}_p{}' = \sum_{p=1}^{N} m_p \ddot{\mathbf{r}}_p. \tag{72}$$

Now, $\sum_{p=1}^{N} \mathbf{F}_p{}' = 0$, since action and reaction between any two particles are equal and opposite, and thus the sum of the internal forces between any pair is zero. Also, the vector position $\mathbf{R}$ of the c.g. of the system is given by $M\mathbf{R} = \sum_{p=1}^{N} m_p \mathbf{r}_p$ (see eqn. (35)) where $M = \sum_{p=1}^{N} m_p$ and hence $M\ddot{\mathbf{R}} = \sum_{p=1}^{N} m_p \ddot{\mathbf{r}}_p$. Substituting from these expressions into eqn. (72) yields

$$\mathbf{T} = M\ddot{\mathbf{R}}$$

where $\mathbf{T}(= \sum_{p=1}^{N} \mathbf{F}_p)$ is the total external force. This is the required result.

(*B*) Taking the vector product of both sides of eqn. (71) with $\mathbf{r}_p$ gives

$$\mathbf{r}_p \wedge F_p + \mathbf{r}_p \wedge F_p{}' = \mathbf{r}_p \wedge m_p \ddot{\mathbf{r}}_p = d(\mathbf{r}_p \wedge m_p \dot{\mathbf{r}}_p)/dt,$$

since

$$d(\mathbf{r}_p \wedge m_p \dot{\mathbf{r}}_p)/dt = \mathbf{r}_p \wedge m_p \ddot{\mathbf{r}}_p + \dot{\mathbf{r}}_p \wedge m_p \dot{\mathbf{r}}_p$$

and the second term is zero.

E

Therefore

$$\sum_p \mathbf{r}_p \wedge \mathbf{F}_p + \sum_p \mathbf{r}_p \wedge \mathbf{F}_p' = d[\sum_p (\mathbf{r}_p \wedge m_p \dot{\mathbf{r}}_p)]/dt. \quad (73)$$

Further, $\sum_p \mathbf{r}_p \wedge \mathbf{F}_p = \mathbf{M}$ the total moment of the external forces and $\sum_p \mathbf{r}_p \wedge \mathbf{F}_p' = 0$ since action and reaction are equal and opposite. Since the total angular momentum $\mathbf{P}$ of the system is given by $\mathbf{P} = \sum_p (\mathbf{r}_p \wedge m_p \dot{\mathbf{r}}_p)$, eqn. (73) yields

$$\mathbf{M} = d\mathbf{P}/dt,$$

the required result.

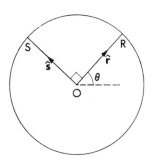

FIG. 32.

Exercises

(1) A particle moves along a plane curve whose polar coordinates (r, θ) are given in terms of the time t by $r = 2t^2 - t$, $\theta = 4t$. What is the magnitude of the acceleration when $t = 2$?

(2) R and S are two points on the circumference of a circle of unit radius as shown in Fig. 32. If $\overrightarrow{OR} = \hat{\mathbf{r}}$ and $\overrightarrow{OS} = \hat{\mathbf{s}}$, show from first principles that if θ varies with time and OR is always perpendicular to OS, then $\dot{\hat{\mathbf{r}}} = \dot{\theta}\hat{\mathbf{s}}$ and $\dot{\hat{\mathbf{s}}} = -\dot{\theta}\hat{\mathbf{r}}$. Hence, by letting $\mathbf{r} = r\hat{\mathbf{r}}$ obtain the radial and transverse components of acceleration for general motion in two dimensions.

(3) A particle P is acted upon by a force which is always directed towards a fixed point O. Show that $\overrightarrow{OP} \wedge \mathbf{v} = \mathbf{h}$ where $\mathbf{v}$ is the particle's velocity and $\mathbf{h}$ is a constant vector.

4.5   INTEGRATION OF VECTORS

We proceed to define the integral of a vector with respect to a scalar as the limit of a sum, and then show that this definition is equivalent to inverting the differentiation process.

Consider a vector $\mathbf{G} = \mathbf{G}(u)$ and suppose u to vary over the range a to b; i.e. $a \leqq u \leqq b$. This range of u is divided into N parts by means of the points $u_1, u_2, u_3, \ldots, u_p, \ldots, u_{N-1}$ as shown in Fig. 33 giving rise to N intervals for u; i.e. $a \leqq u \leqq u_1$,

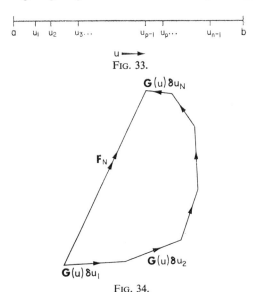

FIG. 33.

FIG. 34.

$u_1 \leqq u \leqq u_2, \ldots, u_{p-1} \leqq u \leqq u_p, \ldots, u_{N-1} \leqq u \leqq b$. For any such interval we define $\delta u_p = u_p - u_{p-1}$, together with the vector $\mathbf{H}_p = \mathbf{G}(u)\delta u_p$, where $\mathbf{G}$ is evaluated at some u lying in the corresponding interval $u_{p-1} \leqq u \leqq u_p$. We then proceed to evaluate the vector $\mathbf{F}_N = \sum_{p=1}^{N} \mathbf{H}_p = \sum_{p=1}^{N} \mathbf{G}(u)\delta u_p$, the summation being shown geometrically in Fig. 34. Now let $N \to \infty$ and

$\delta u_p \to 0$. If $\mathbf{F}_N$ has a definite limiting value $\mathbf{F}$ as $N \to \infty$ independent of the mode of division of the range of u, then the integral of $\mathbf{G}(u)$ over the range a to b is defined as this limiting value; i.e.

$$\mathbf{F} = \int_a^b \mathbf{G}(u)\, du = \lim_{\delta u_p \to 0} \sum_{p=1}^N \mathbf{G}(u)\, \delta u_p. \qquad (74)$$

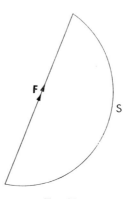

FIG. 35.

The geometrical interpretation of the integral is shown in Fig. 35, the smooth curve S being the limit of the polygon in Fig. 34.

Now suppose that the upper limit b of the integral is variable and therefore $\mathbf{F} = \mathbf{F}(b)$. We have

$$\frac{d\mathbf{F}}{db} = \lim_{\delta b \to 0} \frac{\mathbf{F}(b + \delta b) - \mathbf{F}(b)}{\delta b} = \lim_{\delta b \to 0} \frac{\int_a^{b + \delta b} \mathbf{G}\, du - \int_a^b \mathbf{G}\, du}{\delta b}$$

$$= \lim_{\delta b \to 0} \int_b^{b + \delta b} \mathbf{G}\, du / \delta b. \qquad (75)$$

For δb small, $\mathbf{G}$ is approximately constant over the range

$b \leqq u \leqq b + \delta b$, and thus. $\int_b^{b+\delta b} \mathbf{G}\, \mathrm{d}u = \delta b\mathbf{G}(u)$ for some u in this range. Hence from eqn. (75)

$$\mathrm{d}\mathbf{F}/\mathrm{d}b = \lim_{\delta b \to 0} \delta b\mathbf{G}(u)/\delta b = \lim_{\delta b \to 0} \mathbf{G}(u).$$

We see then that as $\delta b \to 0$, $u \to b$ and so in the limit

$$\mathrm{d}\mathbf{F}/\mathrm{d}b = \mathbf{G}(b).$$

This proves that the above definition of integration is equivalent to inverting the differentiation process. We can then deduce immediately, from the analogous result concerning differentiation that

$$\int \mathbf{G}(u)\, \mathrm{d}u = \mathbf{i} \int G_x(u)\, \mathrm{d}u + \mathbf{j} \int G_y(u)\, \mathrm{d}u + \mathbf{k} \int G_z(u)\, \mathrm{d}u. \quad (76)$$

Worked examples

(1)   If $\mathbf{F} = u^2\mathbf{i} + 2u\mathbf{j} - (u + 1)\mathbf{k}$ and
$$\mathbf{G} = -2\mathbf{i} + 3u\mathbf{j} + (u^2 - 1)\mathbf{k},$$

evaluate $\int_0^2 (\mathbf{F} \wedge \mathbf{G})\, \mathrm{d}u$.

Ans. $\mathbf{F} \wedge \mathbf{G} = \mathbf{i}(2u \times \overline{u^2 - 1} - \overline{u + 1} \times 3u) +$

$$+ \mathbf{j}(\overline{u + 1} \times 2 - u^2 \times \overline{u^2 - 1}) + \mathbf{k}(u^2 \times 3u + 2 \times 2u)$$
$$= \mathbf{i}(2u^3 - 3u^2 - 5u) + \mathbf{j}(-u^4 + u^2 + 2u + 2) +$$
$$+ \mathbf{k}(3u^3 + 4u).$$

Therefore

$$\int_0^2 (\mathbf{F} \wedge \mathbf{G})\, \mathrm{d}u$$

$$= \mathbf{i} \int_0^2 (2u^3 - 3u^2 - 5u)\, \mathrm{d}u + \mathbf{j} \int_0^4 (-u^4 + u^2 + 2u + 2)\, \mathrm{d}u +$$

$$+ \mathbf{k} \int_0^2 (3u^3 + 4u)\, \mathrm{d}u$$

$$= \mathbf{i}(\tfrac{1}{2} \times 2^4 - 2^3 - \tfrac{5}{2} \times 2^2) +$$
$$+ \mathbf{j}(-\tfrac{1}{5} \times 2^5 + \tfrac{1}{3} \times 2^3 + 2^2 + 2 \times 2) + \mathbf{k}(\tfrac{3}{4} \times 2^4 + 2 \times 2^2)$$
$$= -10\mathbf{i} + 4\tfrac{4}{15}\mathbf{j} + 20\mathbf{k}.$$

(2) If $\ddot{\mathbf{r}} = -n^2\mathbf{r}$ for constant n, find $|\dot{\mathbf{r}}|$ as a function of r.

Ans. Taking the scalar product of both sides of the equation with $2\dot{\mathbf{r}}$ gives $2\dot{\mathbf{r}}.\ddot{\mathbf{r}} = -2n^2\dot{\mathbf{r}}.\mathbf{r}$ and therefore

$$d(\dot{\mathbf{r}}.\dot{\mathbf{r}})/dt = -n^2 d(\mathbf{r}.\mathbf{r})/dt$$

making use of the results of §4.2. Integrating both sides of this equation with respect to t yields $\dot{\mathbf{r}}.\dot{\mathbf{r}} = -n^2\mathbf{r}.\mathbf{r} + c$ for arbitrary constant c, and thus $|\dot{\mathbf{r}}| = (c - n^2 r^2)^{\frac{1}{2}}$.

Exercises

(1) If $\mathbf{F} = (u^2 + 1)\mathbf{i} - u\mathbf{j} - (u^2 + 2u)\mathbf{k}$, $\mathbf{G} = 2u\mathbf{i} + (u^2 + 3u)\mathbf{j} - u^3\mathbf{k}$, $\mathbf{H} = 4\mathbf{i} + (2u^2 + 3)\mathbf{j} - u^2\mathbf{k}$, evaluate (*a*) $\int_1^2 \mathbf{F}.\mathbf{G}\,du$, (*b*) $\int_0^3 \mathbf{F} \wedge (\mathbf{G} \wedge \mathbf{H})\,du$, (*c*) $\int_0^2 [\mathbf{F} \wedge (d\mathbf{H}/du)]\,du$, (*d*) $\int_0^2 [\mathbf{H} \wedge (d\mathbf{F}/du)]\,du$. Verify that $(c) - (d) = (\mathbf{F} \wedge \mathbf{H})_{u=2} - (\mathbf{F} \wedge \mathbf{H})_{u=0}$ and prove that this result is true for any two vectors $\mathbf{F}$ and $\mathbf{H}$.

(2) Prove that for any vector $\mathbf{G}(u)$, and scalars a, b,

$$\left| \int_a^b \mathbf{G}(u)\,du \right| \leqq \int_a^b |\mathbf{G}(u)|\,du.$$

4.6 PARTIAL DIFFERENTIATION

In §4.1 we defined the derivative of a vector $\mathbf{F}$ which was a function of a single independent scalar u. A vector may, however, depend on more than one independent scalar in just the same way as in ordinary calculus a quantity z may be a function of two independent variables x and y. When this is the case, we can define partial derivatives of the vector with respect to each of the independent variables. For example, if $\mathbf{F}$ is a function of the two independent scalars u and v, we write $\mathbf{F} = \mathbf{F}(u, v)$ and define

$$\frac{\partial \mathbf{F}}{\partial u} = \lim_{\delta u \to 0} \frac{\mathbf{F}(u + \delta u, v) - \mathbf{F}(u, v)}{\delta u} \tag{77}$$

$$\frac{\partial \mathbf{F}}{\partial v} = \lim_{\delta v \to 0} \frac{\mathbf{F}(u, v + \delta v) - \mathbf{F}(u, v)}{\delta v} \tag{78}$$

Higher derivatives may similarly be defined as in calculus:

$$\frac{\partial^2 \mathbf{F}}{\partial u^2} = \frac{\partial}{\partial u}\left(\frac{\partial \mathbf{F}}{\partial u}\right), \quad \frac{\partial^2 \mathbf{F}}{\partial u \partial v} = \frac{\partial}{\partial u}\left(\frac{\partial \mathbf{F}}{\partial v}\right), \text{ etc.,}$$

and for well behaved functions $\dfrac{\partial^2 \mathbf{F}}{\partial u \partial v} = \dfrac{\partial^2 \mathbf{F}}{\partial v \partial u}$.

The rules in §4.2 for differentiation of sums and products may be easily shown to remain true for partial differentiation. Thus, for example, $\partial(\mathbf{F} \wedge \mathbf{G})/\partial u = (\partial \mathbf{F}/\partial u) \wedge \mathbf{G} + \mathbf{F} \wedge (\partial \mathbf{G}/\partial u)$,

$$\frac{\partial^2 (\mathbf{F}.\mathbf{G})}{\partial u \partial v} = \frac{\partial}{\partial u}\left[\frac{\partial(\mathbf{F}.\mathbf{G})}{\partial v}\right] = \frac{\partial}{\partial u}\left[\frac{\partial \mathbf{F}}{\partial v} \cdot \mathbf{G} + \mathbf{F} \cdot \frac{\partial \mathbf{G}}{\partial v}\right]$$

$$= \frac{\partial^2 \mathbf{F}}{\partial u \partial v} \cdot \mathbf{G} + \frac{\partial \mathbf{F}}{\partial v} \cdot \frac{\partial \mathbf{G}}{\partial u} + \frac{\partial \mathbf{F}}{\partial u} \cdot \frac{\partial \mathbf{G}}{\partial v} + \mathbf{F} \cdot \frac{\partial^2 \mathbf{G}}{\partial u \partial v}.$$

Similarly, the effect of a partial differential operator acting on a vector is given by the same operator acting on its components; for example

$$\frac{\partial^2 \mathbf{F}}{\partial u \partial v} = \mathbf{i}\,\frac{\partial^2 F_x}{\partial u \partial v} + \mathbf{j}\,\frac{\partial^2 F_y}{\partial u \partial v} + \mathbf{k}\,\frac{\partial^2 F_z}{\partial u \partial v}.$$

Finally, by considering the components of each side of the equation, it may be readily shown that for small changes du, dv, in u and v

$$d\mathbf{F} = \frac{\partial \mathbf{F}}{\partial u}\,du + \frac{\partial \mathbf{F}}{\partial v}\,dv. \tag{79}$$

This is, of course, the vector analogue of the scalar equation

$$dz = (\partial z/\partial x)\,dx + (\partial z/\partial y)\,dy$$

for $z = z(x, y)$.

Worked example

If $\mathbf{F} = (3uv^2 + 4u)\mathbf{i} + (u^2v - v^3)\mathbf{j} + (u^4 - v^4)\mathbf{k}$, evaluate $\partial^2\mathbf{F}/\partial u \partial v$ at $u = 2$, $v = 1$.

Ans. $\partial\mathbf{F}/\partial v = 6uv\mathbf{i} + (u^2 - 3v^2)\mathbf{j} - 4v^3\mathbf{k}$,

$$\partial^2\mathbf{F}/\partial u \partial v = 6v\mathbf{i} + 2u\mathbf{j},$$

and therefore at $u = 2$, $v = 1$,

$$\partial^2\mathbf{F}/\partial u \partial v = 6\mathbf{i} + 4\mathbf{j}.$$

Exercises

(1) By expressing $\mathbf{F}.\mathbf{G}$ in terms of the resolutes of $\mathbf{F}$ and $\mathbf{G}$, prove the formula for $\partial^2(\mathbf{F}.\mathbf{G})/\partial u \, \partial v$ given in this section.

(2) If $\mathbf{F}$ is a function of u and v, each of which is itself a function of t, prove the formula

$$\frac{d\mathbf{F}}{dt} = \frac{\partial\mathbf{F}}{\partial u}\frac{du}{dt} + \frac{\partial\mathbf{F}}{\partial v}\frac{dv}{dt}$$

by expressing each side in component form.

(3) If $\mathbf{F} = 4e^{3uv}\mathbf{i} + (u^3v - uv^3)\mathbf{j} + u^2 \sin v\mathbf{k}$, evaluate $\partial\mathbf{F}/\partial u$, $\partial\mathbf{F}/\partial v$, $\partial^2\mathbf{F}/\partial u^2$, $\partial^2\mathbf{F}/\partial v^2$, $\partial^2\mathbf{F}/\partial u \, \partial v$, $\partial^2\mathbf{F}/\partial v \partial u$, at $u = 2$, $v = 1$.

(4) If $\mathbf{F} = [\mathbf{G} \exp i\omega(u - v)]/v$ for a constant vector $\mathbf{G}$, show that

$$\frac{\partial^2\mathbf{F}}{\partial v^2} + \frac{2}{v}\frac{\partial\mathbf{F}}{\partial v} = \frac{\partial^2\mathbf{F}}{\partial u^2}.$$

5

Gradient, Divergence and Curl

5.1 VECTOR AND SCALAR FIELDS

In §4.6 we considered vector functions of two independent variables, together with the corresponding derivatives. In this and succeeding chapters we shall be interested in vectors and scalars which depend on position in three dimensions; i.e. which are functions of three independent space variables such as the cartesian coordinates x, y, z. Given a set of three-dimensional mutually orthogonal cartesian axes OX, OY, OZ, as shown in Fig. 36, the position of any point P in space may be determined by allotting to it cartesian coordinates x, y, z. If then there exists a vector $\mathbf{F}$ which varies with position, this may be denoted by considering $\mathbf{F}$ to be a function of x, y and z, i.e. by putting $\mathbf{F} = \mathbf{F}(x, y, z)$. Since

$$\overrightarrow{OP} = \mathbf{r} = \mathbf{i}x + \mathbf{j}y + \mathbf{k}z$$

this relationship may alternatively be denoted by $\mathbf{F} = \mathbf{F}(\mathbf{r})$, which means that for every position vector $\mathbf{r}$ there exists a corresponding vector $\mathbf{F}$. It should be clear that a vector relationship of the form $\mathbf{F} = \mathbf{F}(\mathbf{r})$ can be looked upon as the vector analogue of the scalar relationship $y = y(x)$ met with in elementary algebra. Expressing $\mathbf{F}$ in component form, i.e. $\mathbf{F} = \mathbf{i}F_x + \mathbf{j}F_y + \mathbf{k}F_z$, it is clear that $F_x = F_x(x, y, z) = F_x(\mathbf{r})$, $F_y = F_y(x, y, z) = F_y(\mathbf{r})$ and $F_z = F_z(x, y, z) = F_z(\mathbf{r})$, where the use of $\mathbf{r}$ as independent variable again means that to every position vector $\mathbf{r}$ there corresponds a value of F_x,

F_y, F_z. A vector **F** depending on position as described here is said to constitute a *vector field*. The specification of such a field is often given via the resolutes of **F**, a typical field being, for example,

$$\mathbf{F} = x^3 e^y \mathbf{i} + (x^2 + y^2 - z^2)\mathbf{j} + z^2 \sin xy\mathbf{k},$$

where $F_x = x^3 e^y$, $F_y = x^2 + y^2 - z^2$, $F_z = z^2 \sin xy$,

and where $F = [x^6 e^{2y} + (x^2 + y^2 - z^2)^2 + z^4 \sin^2 xy]^{\frac{1}{2}}$.

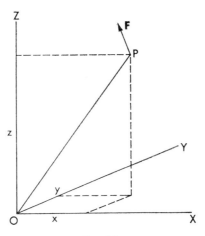

Fig. 36.

Scalar Fields

A scalar quantity V is said to constitute a *scalar field* if it depends on position; i.e. if $V = V(x, y, z) = V(\mathbf{r})$. An example of such a field would be $V(\mathbf{r}) = x^2 + yze^z$.

Although cartesian coordinates are the most useful, for many purposes other coordinate systems are employed. Since we shall be referring to them later, it is convenient to consider two such systems now: (1) spherical polar coordinates, and (2) cylindrical polar coordinates.

(1) *Spherical Polar Coordinates*

Here the position of a point P is specified by (a) the distance OP from a fixed origin O—denoted by r, (b) the angle that OP makes with a fixed direction OZ—denoted by θ, (c) the angle that the plane OPZ makes with a fixed plane OXZ—denoted by φ. These coordinates are shown in Fig. 37, from

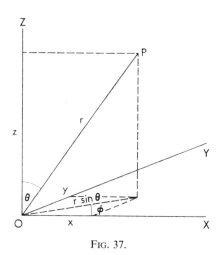

FIG. 37.

which it is seen that they are connected with the cartesian coordinates (x, y, z) via the relations

$$x = r \sin \theta \cos \varphi, \quad y = r \sin \theta \sin \varphi, \quad z = r \cos \theta \quad (80)$$

and conversely

$$r = (x^2 + y^2 + z^2)^{\frac{1}{2}}, \quad \theta = \tan^{-1}[(x^2 + y^2)^{\frac{1}{2}}/z],$$
$$\varphi = \tan^{-1}(y/x).$$

(2) *Cylindrical Polar Coordinates*

Here the position of a point P is specified by (*a*) the distance OS from a fixed origin O to the foot of the perpendicular from P on to a fixed plane XOY—denoted by ρ, (*b*) the angle between OS and OX—denoted by φ, (*c*) the height of P above the plane XOY—denoted by z. These coordinates are shown

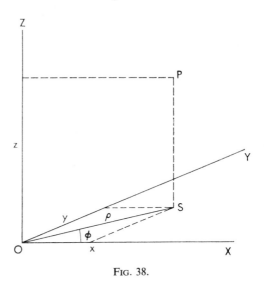

Fig. 38.

in Fig. 38, from which it is seen that they are connected with the cartesian coordinates (x, y, z) via the relation

$$x = \rho \cos \phi, \quad y = \rho \sin \phi, \quad z = z \tag{81}$$

and conversely

$$\rho = (x^2 + y^2)^{\frac{1}{2}}, \quad \varphi = \tan^{-1}(y/x), \quad z = z.$$

It is clear that these coordinates are essentially two-dimensional polar coordinates, together with an additional z coordinate.

As will be seen later, these non-cartesian coordinate systems are of use in physical contexts where the physical situation itself shows certain symmetry—for example, spherical or cylindrical in the two cases dealt with above. In the case of spherical polar coordinates, vector and scalar fields are denoted by $F(r, \theta, \varphi)$, $V(r, \theta, \varphi)$, respectively, while for cylindrical polar coordinates the notation is $F(\rho, \varphi, z)$, $V(\rho, \varphi, z)$.

5.2 THE GRADIENT OPERATOR

It is clear from the general discussions of §4.6 and §5.1 that for a field **F** (vector or scalar) there exist independent first-order partial differential coefficients $\partial F/\partial x$, $\partial F/\partial y$, $\partial F/\partial z$, obtained by acting on **F** with the operators $\partial/\partial x$, $\partial/\partial y$, $\partial/\partial z$ respectively. Now, for reasons which will be discussed in detail in §5.7, it transpires that in many contexts one is interested in certain specific linear combinations of these partial differential coefficients. The first such combination is given by the *gradient* or *grad* operator, defined for a *scalar* field V by

$$\boxed{\text{grad } V = \mathbf{i}(\partial V/\partial x) + \mathbf{j}(\partial V/\partial y) + \mathbf{k}(\partial V/\partial z).} \qquad (82)$$

It is clear that the effect of the grad operator acting on a scalar field V is to yield a vector field **F** given by $F_x = \partial V/\partial x$, $F_y = \partial V/\partial y$, $F_z = \partial V/\partial z$. If we define a vector differential operator ∇ (called " del " or " nabla ") by

$$\boxed{\nabla = \mathbf{i}(\partial/\partial x) + \mathbf{j}(\partial/\partial y) + \mathbf{k}(\partial/\partial z),} \qquad (83)$$

then it follows from eqn. (82) that

$$\text{grad } V = \nabla V. \qquad (84)$$

Small Changes in V

Consider now the change in V corresponding to a small change dx, dy, dz in x, y, z respectively. It is a basic result in partial differentiation theory that

$$dV = \frac{\partial V}{\partial x}dx + \frac{\partial V}{\partial y}dy + \frac{\partial V}{\partial z}dz. \tag{85}$$

Further, the displacement vector $d\mathbf{r}$ corresponding to these changes in x, y and z is given by

$$d\mathbf{r} = \mathbf{i}\,dx + \mathbf{j}\,dy + \mathbf{k}\,dz. \tag{86}$$

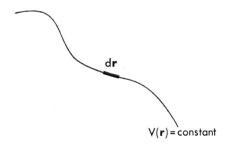

FIG. 39.

It follows then from eqns. (83), (85) and (86) that

$$\boxed{dV = \operatorname{grad} V . d\mathbf{r}.} \tag{87}$$

Suppose now that we consider the equation

$$V(\mathbf{r}) = V(x, y, z) = \text{constant.}$$

This will, in general, represent a surface in three dimensions, the form of the surface for given $V(\mathbf{r})$ depending on the value of the constant. Thus if the constant takes a set of different values, a family of surfaces will be generated. Let us focus attention on a particular surface of this family corresponding to a particular value of the constant and let d**r** in eqn. (87) correspond to a displacement *on* this surface as shown in Fig. 39. Then d$V = 0$ since V is constant over the surface, and thus grad V.d$r = 0$ for *all* such small displacements on the surface. As neither grad V nor d**r** is zero, it follows that grad V is perpendicular to d**r**; i.e. grad V is perpendicular to the surface. Hence the direction of grad V at any point is along the normal at that point to the surface $V(\mathbf{r}) = $ constant.

Rate of Change of V

Take now a general displacement ds in a direction parallel to a given unit vector **t**. Then since d**r** = **t** ds, it follows from eqn. (87) that d$V = $ grad V.**t** ds, and so

$$\mathrm{d}V/\mathrm{d}s = \mathrm{grad}\ V.\mathbf{t}, \tag{88}$$

which is an equation giving the rate at which V is changing with respect to distance measured in any direction. In particular, d$V/$ds will be a maximum when **t** is parallel to grad V, and for this direction d$V/$d$s = |\mathrm{grad}\ V|$. This means that grad V is a vector in the direction along which V is changing most rapidly, and the magnitude of grad V is this most rapid rate of change. Comparing this result with that derived at the end of the last paragraph, it follows that the most rapid rate of change of V is along the normal to the surface $V(\mathbf{r}) = $ constant; this can, of course, be shown directly. If the displacement ds is in a direction with direction cosines (l, m, n) [defined in eqn. (21)], then $\mathbf{t} = l\mathbf{i} + m\mathbf{j} + n\mathbf{k}$ (since $l^2 + m^2 + n^2 = 1$) and so from eqn. (88) it follows that

$$\mathrm{d}V/\mathrm{d}s = l(\partial V/\partial x) + m(\partial V/\partial y) + n(\partial V/\partial z). \tag{89}$$

It should be noted that the notation " ∇V " is used interchangeably with " grad V ", so that eqns. (87) and (88) could be written d$V = \nabla V$.d**r** and d$V/$d$s = \nabla V$.**t** respectively.

Sums and Products

It remains to consider simplified forms that can be obtained for the gradients of sums and products. Let U and V be scalar fields. Then

$$\nabla(U + V) = \mathbf{i}[\partial(U + V)/\partial x] + \mathbf{j}[\partial(U + V)/\partial y] + \\ + \mathbf{k}[\partial(U + V)/\partial z]$$

$$= [\mathbf{i}(\partial U/\partial x) + \mathbf{j}(\partial U/\partial y) + \mathbf{k}(\partial U/\partial z)] + \\ + [\mathbf{i}(\partial V/\partial x) + \mathbf{j}(\partial V/\partial y) + \mathbf{k}(\partial V/\partial z)],$$

and therefore $\qquad \nabla(U + V) = \nabla U + \nabla V.$ $\qquad(90)$

$$\nabla(U \times V) = \mathbf{i}[\partial(U \times V)/\partial x] + \mathbf{j}[\partial(U \times V)/\partial y] + \\ + \mathbf{k}[\partial(U \times V)/\partial z]$$

$$= \mathbf{i}[U(\partial V/\partial x) + V(\partial U/\partial x)] + \mathbf{j}[U(\partial V/\partial y) + \\ + V(\partial U/\partial y)] + \mathbf{k}[U(\partial V/\partial z) + V(\partial U/\partial z)]$$

$$= U[\mathbf{i}(\partial V/\partial x) + \mathbf{j}(\partial V/\partial y) + \mathbf{k}(\partial V/\partial z)] + \\ + V[\mathbf{i}(\partial U/\partial x) + \mathbf{j}(\partial V/\partial y) + \mathbf{k}(\partial U/\partial z)],$$

and therefore

$$\nabla(U \times V) = U \times \nabla V + V \times \nabla U.$$ $\qquad(91)$

It should be noted that in eqns. (90) and (91) the behaviour of the ∇ operator is formally identical with the $\mathrm{d}/\mathrm{d}x$ operator of elementary calculus.

Finally, we wish to pose the following question. Can the relation (82) be inverted: i.e. given a vector field $\mathbf{F}$, is it always possible to obtain a scalar field V such that $\mathbf{F} = \mathrm{grad}\ V$? This question will be considered in detail in §5.6 where it will be shown that in general it is not possible to find V; only if $\mathbf{F}$ satisfies certain conditions, does V exist.

Worked examples

(1) If $V = x^2 \sin y + y\mathrm{e}^z$, find grad V and $\left|\text{grad } V\right|$ at the point $(3, 2, 1)$.

Ans.

$$\partial V/\partial x = 2x \sin y, \quad \partial V/\partial y = x^2 \cos y + \mathrm{e}^z, \quad \partial V/\partial z = y\mathrm{e}^z,$$

and therefore

$$\text{grad } V = \mathbf{i}2x \sin y + (x^2 \cos y + \mathrm{e}^z)\mathbf{j} + y\mathrm{e}^z\mathbf{k}.$$

Therefore at $(3, 2, 1)$,

$$\text{grad } V = \mathbf{i}6 \sin 2 + (9 \cos 2 + \mathrm{e})\mathbf{j} + 2\mathrm{e}\mathbf{k}$$

and $\quad \left|\text{grad } V\right| = (36 \sin^2 2 + \overline{9 \cos 2 + \mathrm{e}^2} + 4\mathrm{e}^2)^{\frac{1}{2}}$

$$= (36 + 45 \cos^2 2 + 18\mathrm{e} \cos 2 + 5\mathrm{e}^2)^{\frac{1}{2}}.$$

(2) If $V = x^2yz + y^3 - z^2$, find the derivative of V at the point $(2, 1, 3)$ with respect to distance measured parallel to $2\mathbf{i} - 2\mathbf{j} + \mathbf{k}$.

Ans. From eqn. (88), $\mathrm{d}V/\mathrm{d}s = \text{grad } V.\mathbf{t}$. Here

$$\text{grad } V = 2xyz\mathbf{i} + (x^2z + 3y^2)\mathbf{j} + (x^2y - 2z)\mathbf{k}$$

$$= 12\mathbf{i} + 15\mathbf{y} - 2\mathbf{k}$$

at the point $(2, 1, 3)$.

Also $\mathbf{t} = (2\mathbf{i} - 2\mathbf{j} + \mathbf{k})/(2^2 + 2^2 + 1)^{\frac{1}{2}} = (1/3)(2\mathbf{i} - 2\mathbf{j} + \mathbf{k}).$

Therefore

$$\mathrm{d}V/\mathrm{d}s = (1/3)(12\mathbf{i} + 15\mathbf{j} - 2\mathbf{k}).(2\mathbf{i} - 2\mathbf{j} + \mathbf{k});$$

i.e. $\mathrm{d}V/\mathrm{d}s = (1/3)(12 \times 2 - 15 \times 2 - 2 \times 1) = -8/3.$

(3) Find unit vectors normal to the surfaces $x^2 - y^2 - z^2 = 11$ and $xy + yz - xz = 18$ at the point $(6, 4, 3)$ and hence find the angle between the two surfaces at this point.

F

Ans. Since grad V is normal to the surface $V(x, y, z)$ = constant, a vector normal to the surface $x^2 - y^2 - z^2 = 11$ is $\nabla(x^2 - y^2 - z^2) = 2x\mathbf{i} - 2y\mathbf{j} - 2z\mathbf{k} = 12\mathbf{i} - 8\mathbf{j} - 6\mathbf{k}$ at the point $(6, 4, 3)$. Thus the required unit vector normal is $\mathbf{p} = (12\mathbf{i} - 8\mathbf{j} - 6\mathbf{k})/(12^2 + 8^2 + 6^2)^{\frac{1}{2}} = (61)^{-\frac{1}{2}}(6\mathbf{i} - 4\mathbf{j} - 3\mathbf{k})$.

Similarly,

$$\nabla(xy + yz - xz) = \mathbf{i}(y - z) + \mathbf{j}(x + z) + \mathbf{k}(y - x) = \mathbf{i} + 9\mathbf{j} - 2\mathbf{k}$$

at $(6, 4, 3)$, and so the unit vector normal at this point to $xy + yz - xz = 18$ is

$$\mathbf{q} = (\mathbf{i} + 9\mathbf{j} - 2\mathbf{k})/(1 + 9^2 + 2^2) = (86)^{-\frac{1}{2}}(\mathbf{i} + 9\mathbf{j} - 2\mathbf{k}).$$

The angle between the surfaces is the angle between the respective unit normals $\mathbf{p}$ and $\mathbf{q}$, and this is

$$\cos^{-1}(\mathbf{p} \cdot \mathbf{q}) = \cos^{-1}[(6 \times 1 - 4 \times 9 + 2 \times 3)/61^{\frac{1}{2}} \times 86^{\frac{1}{2}}] = 109°\ 21'.$$

(4) If $V(\mathbf{r}) = V(r)$, prove grad $V = r^{-1}(\mathrm{d}V/\mathrm{d}r)\mathbf{r}$.

Ans. Since $V(\mathbf{r}) = V(r)$, it follows from eqn. (82) that

$$\text{grad } V = \mathbf{i}\frac{\mathrm{d}V}{\mathrm{d}r}\frac{\partial r}{\partial x} + \mathbf{j}\frac{\mathrm{d}V}{\mathrm{d}r}\frac{\partial r}{\partial y} + \mathbf{k}\frac{\mathrm{d}V}{\mathrm{d}r}\frac{\partial r}{\partial z}$$

$$= \frac{\mathrm{d}V}{\mathrm{d}r}\left(\mathbf{i}\frac{\partial r}{\partial x} + \mathbf{j}\frac{\partial r}{\partial y} + \mathbf{k}\frac{\partial r}{\partial z}\right). \tag{92}$$

Also, $r^2 = x^2 + y^2 + z^2$, whence $2r(\partial r/\partial x) = 2x$. Thus $\partial r/\partial x = x/r$ and similarly $\partial r/\partial y = y/r$, $\partial r/\partial z = z/r$. Substituting these values into eqn. (92) yields

$$\text{grad } V = \frac{\mathrm{d}V}{\mathrm{d}r}\left(\mathbf{i}\frac{x}{r} + \mathbf{j}\frac{y}{r} + \mathbf{k}\frac{z}{r}\right) = \frac{1}{r}\frac{\mathrm{d}V}{\mathrm{d}r}\mathbf{r}.$$

An alternative geometrical derivation of this result is given by employing the fact that grad V is a vector perpendicular to the surface $V(\mathbf{r}) = $ constant, and of magnitude equal to the derivative of V in this direction. Now, if $V(\mathbf{r}) = V(r)$, the surfaces $V(r)$ = constant are spheres centre O and radius r. Hence, since $\mathbf{r}$ at any point is perpendicular to the surface of the sphere passing through that point, it follows that grad V is parallel to $\mathbf{r}$, and $|\text{grad } V| = dV/dr$. Therefore grad V $= r^{-1}(dV/dr)\mathbf{r}$.

(5) Find the vector and cartesian equations of the tangent plane to the surface $3y^2z - x^3y + 4xz = 10$ at the point $(-1, 2, 1)$.

Ans. The normal to the surface is parallel to

$$\mathbf{F} = \text{grad } (3y^2z - x^3y + 4xz) = (4z - 3x^2y)\mathbf{i} + (6yz - x^3)\mathbf{j} +$$

$$+ (3y^2 + 4x)\mathbf{k} = -2i + 13\mathbf{j} + 8\mathbf{k}$$

at $(-1, 2, 1)$. The tangent plane is perpendicular to the normal, and thus if $\overrightarrow{OP} = \mathbf{r}$ where P is a point in the plane, it follows from eqn. (23) that the equation of the plane will be

$$(\mathbf{r} - \mathbf{R}) . \mathbf{F} = 0$$

where S is the point $(-1, 2, 1)$ and $\mathbf{R} = \overrightarrow{OS}$. If P has cartesian coordinates (x, y, z) the equation $(\mathbf{r} - \mathbf{R}) . \mathbf{F}$ becomes

$$[(x + 1)\mathbf{i} + (y - 2)\mathbf{j} + (z - 1)\mathbf{k}] . (-2\mathbf{i} + 13\mathbf{j} + 8\mathbf{k}) = 0;$$

i.e. $-2(x + 1) + 13(y - 2) + 8(z - 1) = 0$. The required cartesian equation is thus $2x - 13y - 8z + 36 = 0$.

(6) If $\mathbf{F}$ and V are vector and scalar fields, show that the components of $\mathbf{F}$ at a point, in directions normal and tangential to the surface of constant V passing through that point, are

$$\frac{(\mathbf{F} . \nabla V)\nabla V}{\nabla V . \nabla V} \quad \text{and} \quad \frac{\nabla V \wedge (\mathbf{F} \wedge \nabla V)}{\nabla V . \nabla V} \text{ respectively.}$$

Ans. ∇V is a vector perpendicular to the surface of constant V as shown in Fig. 39, and therefore the resolute of $\mathbf{F}$ perpendicular to the surface is $F \cos \theta = (\mathbf{F} . \nabla V)/|\nabla V|$.

Hence the component of $\mathbf{F}$ perpendicular to the surface is

$$\frac{(\mathbf{F}.\nabla V)}{|\nabla V|} \times \frac{\nabla V}{|\nabla V|} = \frac{(\mathbf{F}.\nabla V)\nabla V}{\nabla V.\nabla V}.$$

The component of F tangential to the surface is thus

$$\mathbf{F} - \frac{(\mathbf{F}.\nabla V)\nabla V}{\nabla V.\nabla V} = \frac{\mathbf{F}(\nabla V.\nabla V) - (\mathbf{F}.\nabla V)\nabla V}{\nabla V.\nabla V}$$

$$= \frac{\nabla V \wedge (\mathbf{F} \wedge \nabla V)}{\nabla V.\nabla V}.$$

Exercises

(1) If $\mathbf{F} = x^3z\mathbf{i} + (y^2 - z^2)\mathbf{j} + (x^2 + y^2 + z^2)\mathbf{k}$ and $V = y^2z^3 - x^2y + xz$, calculate $\mathbf{F} \wedge \text{grad } V$ at the point $(2, 0, -1)$.
(2) Prove $\nabla(U/V) = (V \times \nabla U - U \times \nabla V)/V^2$.
(3) If $V = \sin(x^2 + y^2 + z^2)^{\frac{1}{2}}$ find the derivative of V with respect to distance measured parallel to $3\mathbf{i} - 2\mathbf{j} + \mathbf{k}$ at the point $(-2, 2, 1)$.
(4) If $\mathbf{F}$ is a constant vector, prove that grad $(\mathbf{F}.\mathbf{r}) = \mathbf{F}$.
(5) If $V = ax^2y + bx^3z + cy^2z$ find a, b and c so that the maximum rate of change of V with respect to distance at the point $(3, -1, 2)$ is 27 in a direction parallel to the y axis.
(6) Find the constants a and b so that the surface $ax^3 - by^2z - (a + 3)x^2 = 0$ is perpendicular to the surface $4x^2y - z^3 - 11 = 0$ at the point $(2, -1, -3)$.

5.3   THE DIVERGENCE OPERATOR

In the last section we considered the grad operator which acting on a *scalar* field gave rise to a *vector* field. We consider now the *divergence* or *div* operator which acts on a *vector* field and gives rise to a *scalar* field via the following definition:

$$\boxed{\text{div } \mathbf{F} = (\partial F_x/\partial x) + (\partial F_y/\partial y) + (\partial F_z/\partial z).} \tag{93}$$

Employing the ∇ operator defined in eqn. (83), it is clear that using the usual form for the scalar product of two vectors in terms of their resolutes,

$$\text{div } \mathbf{F} = \nabla.\mathbf{F} \tag{94}$$

since

$$\nabla . \mathbf{F} = [\mathbf{i}(\partial/\partial x) + \mathbf{j}(\partial/\partial y) + \mathbf{k}(\partial/\partial z)] . (\mathbf{i}F_x + \mathbf{j}F_y + \mathbf{k}F_z)$$
$$= (\partial/\partial x)F_x + (\partial/\partial y)F_y + (\partial/\partial z)F_z.$$

As in the last section, the notation " div " or " ∇." may be used interchangeably.

It is readily shown that

$$\nabla .(\mathbf{F} + \mathbf{G}) = \nabla . \mathbf{F} + \nabla . \mathbf{G} \qquad (95)$$

since

$$\nabla .(\mathbf{F} + \mathbf{G}) = \partial(F + G)_x/\partial x + \partial(F + G)_y/\partial y + \partial(F + G)_z/\partial z$$
$$= (\partial F_x/\partial x + \partial F_y/\partial y + \partial F_z/\partial z) +$$
$$+ (\partial G_x/\partial x + \partial G_y/\partial y + \partial G_z/\partial z)$$
$$= \nabla . \mathbf{F} + \nabla . \mathbf{G}.$$

A consideration of the div operator acting on products will be deferred until §5.5, while the significance of the div operator will be discussed after the relevant integration theorem in §7·2.

A vector $\mathbf{F}$, such that div $\mathbf{F} = 0$ everywhere is said to be a *solenoidal* vector.

It is clear that eqn. (93) can readily be inverted; i.e. given a scalar field V it is always possible to find a vector field $\mathbf{F}$ such that div $\mathbf{F} = V$. However, the field $\mathbf{F}$ will not be unique since if $\partial F_x/\partial x = V^{(x)}$, $\partial F_y/\partial y = V^{(y)}$, $\partial F_z/\partial z = V^{(z)}$, then $V^{(x)}$, $V^{(y)}$, $V^{(z)}$ are arbitrary as long as their sum equals V. Further, $F_x = \int V^{(x)}dx$ and this indefinite integral will be undefined to the extent of an additive arbitrary function of y and z. Similar remarks apply to F_y and F_z.

It should be noted that $\nabla . \mathbf{F} \neq \mathbf{F} . \nabla$. The former is div $\mathbf{F}$ which is a scalar, but the latter expression,

$$\mathbf{F} . \nabla = F_x(\partial/\partial x) + F_y(\partial/\partial y) + F_z(\partial/\partial z),$$

which is an *operator*. We shall have occasion to employ this operator in §5.5.

Worked examples

(1) If $\mathbf{F} = (3x^2y - z^3)\mathbf{i} + (x^2y^2 + y^3z)\mathbf{j} + (xz^3 - yz^2)\mathbf{k}$, calculate div $\mathbf{F}$ at the point $(-2, 3, 2)$.

Ans.

$\partial F_x/\partial x = 6xy$, $\partial F_y/\partial y = 3y^2z$, $\partial F_z/\partial z = 3xz^2 - 2yz$, and therefore div $\mathbf{F} = 6xy + 3y^2z + 3xz^2 - 2yz = -18$

at $(-2, 3, 2)$.

(2) If div $\mathbf{F} = 3x^2y - 2xyz + yz^3$ obtain a possible form for $\mathbf{F}$.

Ans. We may take $\partial F_x/\partial x = 3x^2y$, $\partial F_y/\partial y = -2xyz$, $\partial F_z/\partial z = yz^3$, and integrating these gives $F_x = x^3y$, $F_y = -xy^2z$, $F_z = \frac{1}{4}yz^4$, omitting any arbitrary functions. Hence a possible solution for $\mathbf{F}$ is $\mathbf{F} = x^3y\mathbf{i} - xy^2z\mathbf{j} + \frac{1}{4}yz^4\mathbf{k}$.

(3) If $\mathbf{F}(\mathbf{r}) = f(r)\mathbf{r}$, prove that div $\mathbf{F} = 3f + r\,df/dr$.

Ans. We have $\mathbf{F}(\mathbf{r}) = \mathbf{i}f(r)x + \mathbf{j}f(r)y + \mathbf{k}f(r)z$, and therefore

$$\text{div } \mathbf{F} = \partial[xf(r)]/\partial x + \partial[yf(r)]/\partial y + \partial[zf(r)]/\partial z$$

$$= f + x\partial f/\partial x + f + y\partial f/\partial y + f + z\partial f/\partial z$$

$$= 3f + x(df/dr)(\partial r/\partial x) + y(df/dr)(\partial r/\partial y) + z(df/dr)(\partial r/\partial z)$$

$$= 3f + (df/dr)[(x^2/r) + (y^2/r) + (z^2/r)]$$

since $\qquad \dfrac{\partial r}{\partial x} = \dfrac{x}{r}, \quad \dfrac{\partial r}{\partial y} = \dfrac{y}{r}, \quad \dfrac{\partial r}{\partial z} = \dfrac{z}{r}.$

Therefore div $\mathbf{F} = 3f + r\,df/dr$, since $x^2 + y^2 + z^2 = r^2$.

(4) Show that if $\boldsymbol{\omega}$ is a constant vector, then $\mathbf{F} = \boldsymbol{\omega} \wedge \mathbf{r}$ is a solenoidal vector; i.e. div $\mathbf{F} = 0$.

Ans. We have $F_x = \omega_y z - y\omega_z$, $F_y = \omega_z x - z\omega_x$, $F_z = x\omega_y - y\omega_x$. Therefore $\partial F_x/\partial x = 0$, since $\boldsymbol{\omega}$ is constant, and similarly $\partial F_y/\partial y = \partial F_z/\partial z = 0$.

Thus div $\mathbf{F} = 0$.

Exercises

(1) If $\mathbf{F} = e^{xyz}\mathbf{i} + \ln xyz\mathbf{j} + xyz\mathbf{k}$, calculate div $\mathbf{F}$ at the point $(4, 2, 3)$.
(2) Show that $\mathbf{F} = r^{-3}\mathbf{r}$ is a solenoidal vector.
(3) Find the value of a for the vector
$\mathbf{F} = (2x^2y + yz)\mathbf{i} + (xy^2 - xz^2)\mathbf{j} + (a\,xyz - 2x^2y^2)\mathbf{k}$ to be solenoidal.

5.4 THE CURL OPERATOR

The *curl* operator acts on a *vector* field giving rise to another *vector* field via the definition

$$\text{curl } \mathbf{F} = \mathbf{i}\left(\frac{\partial F_z}{\partial y} - \frac{\partial F_y}{\partial z}\right) + \mathbf{j}\left(\frac{\partial F_x}{\partial z} - \frac{\partial F_z}{\partial x}\right) + \\ + \mathbf{k}\left(\frac{\partial F_y}{\partial x} - \frac{\partial F_x}{\partial y}\right). \tag{96}$$

Employing the ∇ operator, we may write

$$\text{curl } \mathbf{F} = \nabla \wedge \mathbf{F} \tag{97}$$

since

$$\nabla \wedge \mathbf{F} = [\mathbf{i}(\partial/\partial x) + \mathbf{j}(\partial/\partial y) + \mathbf{k}(\partial/\partial z)] \wedge [\mathbf{i}F_x + \mathbf{j}F_y + \mathbf{k}F_z]$$

$$= \mathbf{i}[(\partial/\partial y)F_z - (\partial/\partial z)F_y] + \mathbf{j}[(\partial/\partial_z)F_x - (\partial/\partial x)F_z] + \\ + \mathbf{k}[(\partial/\partial x)F_y - (\partial/\partial y)F_x].$$

The notations " curl " or " $\nabla \wedge$ " may be used interchangeably. An alternative form for curl $\mathbf{F}$ in terms of a determinant is given by

$$\text{curl } \mathbf{F} = \begin{vmatrix} \mathbf{i} & \mathbf{j} & \mathbf{k} \\ \partial/\partial x & \partial/\partial y & \partial/\partial z \\ F_x & F_y & F_z \end{vmatrix}; \tag{98}$$

this follows immediately from eqns. (30), (83), (97).

It will be left to the student to show that

$$\nabla \wedge (\mathbf{F} + \mathbf{G}) = \nabla \wedge \mathbf{F} + \nabla \wedge \mathbf{G} \tag{99}$$

by expressing each side in terms of its cartesian components.

The effect of the curl operator acting on products will be dealt with in the next section, while the significance of the curl will be discussed in Chapter 7.

A vector $\mathbf{F}$, such that curl $\mathbf{F} = 0$ everywhere, is said to be a *lamellar* or *irrotational* vector.

Finally, we may ask whether eqn. (96) may be inverted. That is, given a vector field $\mathbf{G}$, is it always possible to find a vector field $\mathbf{F}$ such that $\mathbf{G} = \text{curl } \mathbf{F}$? We shall show later in §5.6 that a necessary and sufficient condition for $\mathbf{F}$ to exist is that $\mathbf{G}$ is a solenoidal vector; i.e. div $\mathbf{G} = 0$.

Worked examples

(1)   If $\mathbf{F} = (x^2 + y^2 + z^2)\mathbf{i} + (x^4 - y^2z^2)\mathbf{j} + xyz\mathbf{k}$, find curl $\mathbf{F}$ at the point $(2, 3, -2)$.

Ans.

$$\partial F_x/\partial y = 2y, \quad \partial F_x/\partial z = 2z, \quad \partial F_y/\partial x = 4x^3,$$

$$\partial F_y/\partial z = -2y^2z, \quad \partial F_z/\partial x = yz, \quad \partial F_z/\partial y = xz.$$

Hence, from eqn. (96)

$$\text{curl } \mathbf{F} = \mathbf{i}(xz + 2y^2z) + \mathbf{j}(2z - yz) + \mathbf{k}(4x^3 - 2y)$$

$$= -40\mathbf{i} + 2\mathbf{j} + 26\mathbf{k}$$

at the point $(2, 3, -2)$.

(2) Show that $f(r)\mathbf{r}$ is an irrotational vector.

Ans. If $\mathbf{F} = f(r)\mathbf{r}$, $\mathbf{F} = \mathbf{i}f(r)x + \mathbf{j}f(r)y + \mathbf{k}f(r)z$, and therefore

$$\partial F_z/\partial y = z(\partial f/\partial y) = z(\mathrm{d}f/\mathrm{d}r)(\partial r/\partial y) = (yz/r)(\mathrm{d}f/\mathrm{d}r)$$

since $\partial r/\partial y = y/r.$

Similarly, it may be shown that

$$\frac{\partial F_y}{\partial z} = \frac{yz}{r}\frac{\mathrm{d}f}{\mathrm{d}r} = \frac{\partial F_z}{\partial y}$$

whence it follows that $(\text{curl } \mathbf{F})_x = (\partial F_z/\partial y) - (\partial F_y/\partial z) = 0$.
In the same way $(\text{curl } \mathbf{F})_y = (\text{curl } \mathbf{F})_z = 0$, so that curl $\mathbf{F} = 0$,
and therefore $\mathbf{F}$ is irrotational.

(3) Prove that if $\mathbf{F} = (xyz)^b(x^a\mathbf{i} + y^a\mathbf{j} + z^a\mathbf{k})$ is an irrotational vector, then either $b = 0$ or $a = -1$.

Ans. Curl $\mathbf{F} = \text{curl } (x^{a+b}y^bz^b\mathbf{i} + x^by^{a+b}z^b\mathbf{j} + x^by^bz^{a+b}\mathbf{k})$

$$= \mathbf{i}\left[\frac{\partial x^by^bz^{a+b}}{\partial y} - \frac{\partial x^by^{a+b}z^b}{\partial z}\right] + \mathbf{j}\left[\frac{\partial x^{a+b}y^bz^b}{\partial z} - \frac{\partial x^by^bz^{a+b}}{\partial x}\right] +$$

$$+ \mathbf{k}\left[\frac{\partial x^by^{a+b}z^b}{\partial x} - \frac{\partial x^{a+b}y^bz^b}{\partial y}\right]$$

$$= b[\mathbf{i}(x^by^{b-1}z^{a+b} - x^by^{a+b}z^{b-1}) + \mathbf{j}(x^{a+b}y^bz^{b-1} - x^{b-1}y^bz^{a+b}) +$$
$$+ \mathbf{k}(x^{b-1}y^{a+b}z^b - x^{a+b}y^{b-1}z^b)].$$

Now, if curl $\mathbf{F} = 0$, either $b = 0$ or each of the components is
separately zero. Equating therefore each of these components
to zero in turn and cancelling by $(xyz)^b$ gives

$$y^{-1}z^a = y^az^{-1}, \quad x^az^{-1} = x^{-1}z^a, \quad x^{-1}y^a = x^ay^{-1}.$$

Clearly these equations are all satisfied by $a = -1$, and thus
curl $\mathbf{F} = 0$ if $b = 0$ or $a = -1$.

Exercises

(1) If $\mathbf{F} = e^{xyz}\mathbf{i} + xy \sin z\mathbf{j} + (x^3 - y^3)\mathbf{k}$ calculate curl $\mathbf{F}$ at the point $(1, -2, 0)$.

(2) Find a, b and c such that the vector field
$$\mathbf{F} = \mathbf{i}(axy + bz^3) + \mathbf{j}(3x^2 - cz) + \mathbf{k}(3xz^2 - y)$$
is irrotational. For these values of a, b, c, what is div $\mathbf{F}$ at the
point $(0, 1, -2)$?

5.5 GRAD, DIV AND CURL OF PRODUCTS

In §5.2 we proved the result (a) $\nabla(UV) = U\nabla V + V\nabla U$.
In this section we shall consider the other possible forms which
may arise when the ∇ operator acts on a product. They are

(b) $\nabla(\mathbf{F} \cdot \mathbf{G})$, (c) $\nabla \cdot (V\mathbf{F})$, (d) $\nabla \cdot (\mathbf{F} \wedge \mathbf{G})$, (e) $\nabla \wedge (V\mathbf{F})$, (f) $\nabla \wedge (\mathbf{F} \wedge \mathbf{G})$, where $\mathbf{F}$, $\mathbf{G}$ are vector fields and V is a scalar field. First we list the expanded forms of these expressions for reference, and then deal with the proofs of them.

> (a) $\operatorname{grad}(UV) = U \operatorname{grad} V + V \operatorname{grad} U.$ (100)
>
> (b) $\operatorname{grad}(\mathbf{F} \cdot \mathbf{G}) = \mathbf{F} \wedge \operatorname{curl} \mathbf{G} + \mathbf{G} \wedge \operatorname{curl} \mathbf{F} +$
> $$(\mathbf{F} \cdot \operatorname{grad})\mathbf{G} + (\mathbf{G} \cdot \operatorname{grad})\mathbf{F}, \quad (101)$$
> where
> $$\mathbf{F} \cdot \operatorname{grad} = \mathbf{F} \cdot \nabla = F_x(\partial/\partial x) + F_y(\partial/\partial y) +$$
> $$+ F_z(\partial/\partial z) - \text{(see the end of §5.3).}$$
> (c) $\operatorname{div}(V\mathbf{F}) = V \operatorname{div} \mathbf{F} + \mathbf{F} \cdot \operatorname{grad} V.$ (102)
>
> (d) $\operatorname{div}(\mathbf{F} \wedge \mathbf{G}) = \mathbf{G} \cdot \operatorname{curl} \mathbf{F} - \mathbf{F} \cdot \operatorname{curl} \mathbf{G}.$ (103)
>
> (e) $\operatorname{curl}(V\mathbf{F}) = V \operatorname{curl} \mathbf{F} + \operatorname{grad} V \wedge \mathbf{F}.$ (104)
>
> (f) $\operatorname{curl}(\mathbf{F} \wedge \mathbf{G}) = \mathbf{F} \operatorname{div} \mathbf{G} - \mathbf{G} \operatorname{div} \mathbf{F} +$
> $$(\mathbf{G} \cdot \operatorname{grad})\mathbf{F} - (\mathbf{F} \cdot \operatorname{grad})\mathbf{G}. \quad (105)$$

These results may all be proved by writing each side in component form and showing the equality of the resultant expressions. For example, in case (b) the x component of the R.H.S. equals

$$F_y \left(\frac{\partial G_y}{\partial x} - \frac{\partial G_x}{\partial y}\right) - F_z \left(\frac{\partial G_x}{\partial z} - \frac{\partial G_z}{\partial x}\right) + G_y \left(\frac{\partial F_y}{\partial x} - \frac{\partial F_x}{\partial y}\right)$$

$$- G_z \left(\frac{\partial F_x}{\partial z} - \frac{\partial F_z}{\partial x}\right) + F_x \frac{\partial G_x}{\partial x} + F_y \frac{\partial G_x}{\partial y} + F_z \frac{\partial G_x}{\partial z} +$$

$$+ G_x \frac{\partial F_x}{\partial x} + G_y \frac{\partial F_x}{\partial y} + G_z \frac{\partial F_x}{\partial z}$$

$$= F_x \frac{\partial G_x}{\partial x} + G_x \frac{\partial F_x}{\partial x} + F_y \frac{\partial G_y}{\partial x} + G_y \frac{\partial F_y}{\partial x} + F_z \frac{\partial G_z}{\partial x} + G_z \frac{\partial F_z}{\partial x}$$

$$= \partial(F_x G_x + F_y G_y + F_z G_z)/\partial x$$

$$= x \text{ component of } \operatorname{grad}(\mathbf{F} \cdot \mathbf{G});$$

a similar proof may then be used for the y and z components. Now it is clear that this method of proof is rather long and tedious when all the components are dealt with. Further, it can only be used to *prove* a given form for the result, and cannot easily be adapted to provide a straightforward procedure for *finding* the result when this is not known. We shall therefore consider now a shorter approach which provides a method for obtaining the required result *ab initio*, and which does not require a direct appeal to the component formulation. The basis of the method is to employ the ∇ operator throughout the calculation, making use of both its operator and vector properties. The approach is aimed at obtaining the vector generalisation of the scalar result,

$$d(uv)/dx = u(dv/dx) + v(du/dx).$$

Since the proof of the method is somewhat more difficult than the previous work, the student may safely omit it if he so wishes, and proceed directly to the rule and examples of its use as given on pages 84–86.

We first notice that all of the forms (a) to (f) may be written $\nabla \otimes (R \otimes S)$ where R and S are either scalar or vector fields, and where $\otimes$ represents ordinary multiplication of a scalar and a vector, or scalar or vector multiplication of two vectors. We now proceed to show that

$$\nabla \otimes (R \otimes S) = \nabla_R \otimes (R \otimes S) + \nabla_S \otimes (R \otimes S), \quad (106)$$

where ∇_R means that the differential operators in the ∇_R expression act only on R in any function which follows it, and similarly ∇_S operates only on S in any subsequent function. The proof of eqn. (106) follows from the fact that the L.H.S. always consists of a sum of terms of the form $h\partial(R_\alpha S_\beta)/\partial\gamma$, where α, β, γ are each x, y or z as the case may be if R and S are both vector fields, and R_α, $S_\beta = R$, S respectively if either is a scalar field; $h = 1$ if the operator is div and h may be $\pm\mathbf{i}$, $\pm\mathbf{j}$, $\pm\mathbf{k}$ if the operator is grad or curl. Now, clearly

$$h\partial(R_\alpha S_\beta)/\partial\gamma = hR_\alpha(\partial S_\beta/\partial\gamma) + hS_\beta(\partial R_\alpha/\partial\gamma)$$

whence it follows that eqn. (106) is true, since the L.H.S. of this equation is a certain sum of terms of the form $h\partial(R_\alpha S_\beta)/\partial\gamma$, while the two terms on the R.H.S. are respectively equal to identical sums of terms of the form $hS_\beta(\partial R_\alpha/\partial\gamma)$ and $hR_\alpha(\partial S_\beta/\partial\gamma)$. The first stage, then, in the simplification of one of the expressions (a) to (f) is to write it in the form given by the R.H.S. of eqn. (106). In order to obtain the required result it is then necessary to rearrange the terms in each of the expressions on the R.H.S. of eqn. (106) so that the ∇_R and ∇_S operators have on their right only terms in R and S respectively, the S and R terms in the two cases respectively being on the left of the corresponding ∇ operator. Since the ∇ operator is itself a vector, this rearrangement may be done using the rules of vector algebra in the manipulation of the ∇, R and S symbols. Finally, the resulting expressions can be interpreted to give the relevant R.H.S. of eqns. (100) to (105). This approach may be summed up in the following rule:

In order to obtain the R.H.S. of each of the eqns. (100) to (105) write the L.H.S. with ∇ notation as the sum of two terms in each of which the ∇ operator acts on only one of the two fields to be differentiated; the one it acts upon may be shown by writing this as a subscript to the ∇ operator. Transform each of these expressions by the rules of vector algebra, *treating ∇ purely as a vector*, so that the terms to be acted upon by ∇ appears on the R.H.S. of ∇ and the other term on its L.H.S. Finally, interpret the result.

We now apply this approach to the cases (a) to (f). Taking first case (c), we have

$$\text{div}\,(V\mathbf{F}) = \nabla.(V\mathbf{F}) = \nabla_V.(V\mathbf{F}) + \nabla_\mathbf{F}.(V\mathbf{F})$$

$$= \mathbf{F}.\nabla_V V + V\nabla_\mathbf{F}.\mathbf{F} \tag{107}$$

(remember that in the last step the ∇ symbol is manipulated purely as a *vector* and its operational significance is temporarily disregarded, so that $\nabla_V \cdot (V\mathbf{F}) = \mathbf{F} \cdot \nabla_V V$ and $\nabla_\mathbf{F} \cdot (V\mathbf{F}) = V\nabla_\mathbf{F} \cdot \mathbf{F}$). Finally, interpreting the R.H.S. of eqn. (107) gives

$$\text{div} (V\mathbf{F}) = \mathbf{F} \cdot \text{grad } V + V \text{ div } \mathbf{F}.$$

Case (*d*)

$$\text{div} (\mathbf{F} \wedge \mathbf{G}) = \nabla \cdot (\mathbf{F} \wedge \mathbf{G}) = \nabla_\mathbf{F} \cdot (\mathbf{F} \wedge \mathbf{G}) + \nabla_\mathbf{G} \cdot (\mathbf{F} \wedge \mathbf{G})$$

$$= \mathbf{G} \cdot (\nabla_\mathbf{F} \wedge \mathbf{F}) - \mathbf{F} \cdot (\nabla_\mathbf{G} \wedge \mathbf{G}), \tag{108}$$

making use of the invariance of the triple scalar product to cyclic interchange of factors. Finally, interpreting the R.H.S. of eqn. (108) gives eqn. (103).

Case (*e*)

$$\text{curl} (V\mathbf{F}) = \nabla \wedge (V\mathbf{F}) = \nabla_V \wedge (V\mathbf{F}) + \nabla_\mathbf{F} \wedge (V\mathbf{F})$$

$$= -\mathbf{F} \wedge (\nabla_V V) + V\nabla_\mathbf{F} \wedge \mathbf{F}$$

$$= -\mathbf{F} \wedge \text{grad } V + V \text{ curl } \mathbf{F}$$

on interpretation.

Case (*f*)

$$\text{curl} (\mathbf{F} \wedge \mathbf{G}) = \nabla \wedge (\mathbf{F} \wedge \mathbf{G}) = \nabla_\mathbf{F} \wedge (\mathbf{F} \wedge \mathbf{G}) + \nabla_\mathbf{G} \wedge (\mathbf{F} \wedge \mathbf{G})$$

$$= (\mathbf{G} \cdot \nabla_\mathbf{F})\mathbf{F} - (\nabla_\mathbf{F} \cdot \mathbf{F})\mathbf{G} + (\nabla_\mathbf{G} \cdot \mathbf{G})\mathbf{F} -$$

$$- (\mathbf{F} \cdot \nabla_\mathbf{G})\mathbf{G} \tag{109}$$

making use of the result (45). Interpreting eqn. (109) then yields eqn. (105).

Case (*b*)

$$\text{grad} (\mathbf{F} \cdot \mathbf{G}) = \nabla(\mathbf{F} \cdot \mathbf{G}) = \nabla_\mathbf{F}(\mathbf{F} \cdot \mathbf{G}) + \nabla_\mathbf{G}(\mathbf{F} \cdot \mathbf{G}). \tag{110}$$

Now, $\mathbf{F} \wedge (\nabla_\mathbf{G} \wedge \mathbf{G}) = \nabla_\mathbf{G}(\mathbf{F} \cdot \mathbf{G}) - (\mathbf{F} \cdot \nabla_\mathbf{G})\mathbf{G}$

and so $\quad \nabla_\mathbf{G}(\mathbf{F} \cdot \mathbf{G}) = \mathbf{F} \wedge (\nabla_\mathbf{G} \wedge \mathbf{G}) + (\mathbf{F} \cdot \nabla_\mathbf{G})\mathbf{G}.$

Similarly

$$\nabla_F(\mathbf{F}.\mathbf{G}) = \mathbf{G} \wedge (\nabla_F \wedge \mathbf{F}) + (\mathbf{G}.\nabla_F)\mathbf{F}.$$

Substituting these forms into eqn. (110) then gives

$$\nabla(\mathbf{F}.\mathbf{G}) = \mathbf{G} \wedge (\nabla_F \wedge \mathbf{F}) + (\mathbf{G}.\nabla_F)\mathbf{F} + \mathbf{F} \wedge (\nabla_G \wedge \mathbf{G}) + (\mathbf{F}.\nabla_G)\mathbf{G},$$

which on interpretation yields eqn. (101).

Case (a)

$$\begin{aligned}
\operatorname{grad}(UV) = \nabla(UV) &= \nabla_U(UV) + \nabla_V(UV) \\
&= V(\nabla_U U) + U(\nabla_V V) \\
&= V \operatorname{grad} U + U \operatorname{grad} V \quad \text{on interpretation.}
\end{aligned}$$

Expressions such as grad $[\mathbf{F}.(\mathbf{G} \wedge \mathbf{H})]$ or curl $[V(\mathbf{G} \wedge \mathbf{H})]$ arising from the ∇ operator acting on a product of three or more quantities, can be expanded by repeated application of the above results (100) to (105).

Worked examples

(1) If $\mathbf{v} = \boldsymbol{\omega} \wedge \mathbf{r}$ for $\boldsymbol{\omega}$ constant, show that curl $\mathbf{v} = 2\boldsymbol{\omega}$.

Ans.   Curl $\mathbf{v} = \operatorname{curl}(\boldsymbol{\omega} \wedge \mathbf{r}) = \boldsymbol{\omega} \operatorname{div} \mathbf{r} - (\boldsymbol{\omega}.\operatorname{grad})\mathbf{r}$ from eqn. (105), since $\boldsymbol{\omega}$ is constant.

Now,

$$\operatorname{div} \mathbf{r} = (\mathrm{d}x/\mathrm{d}x) + (\mathrm{d}y/\mathrm{d}y) + (\mathrm{d}z/\mathrm{d}z) = 3 \text{ and}$$
$$(\boldsymbol{\omega}.\nabla)\mathbf{r} = [\omega_x(\partial/\partial x) + \omega_y(\partial/\partial y) + \omega_z(\partial/\partial z)](\mathbf{i}x + \mathbf{j}y + \mathbf{k}z)$$
$$= \mathbf{i}\omega_x + \mathbf{j}\omega_y + \mathbf{k}\omega_z = \boldsymbol{\omega}.$$

Therefore

$$\operatorname{curl} \mathbf{v} = 3\boldsymbol{\omega} - \boldsymbol{\omega} = 2\boldsymbol{\omega}.$$

(2) Expand curl $(V\mathbf{F} \wedge \mathbf{G})$.

Ans. From eqn. (104),

$$\begin{aligned}
\operatorname{curl}(V\mathbf{F} \wedge \mathbf{G}) &= V \operatorname{curl} \mathbf{F} \wedge \mathbf{G} + \operatorname{grad} V \wedge (\mathbf{F} \wedge \mathbf{G}) \\
&= V\mathbf{F} \operatorname{div} \mathbf{G} - V\mathbf{G} \operatorname{div} \mathbf{F} + V(\mathbf{G}.\nabla)\mathbf{F} - \\
&\quad - V(\mathbf{F}.\nabla)\mathbf{G} + (\nabla V.\mathbf{G})\mathbf{F} - (\nabla V.\mathbf{F})\mathbf{G}
\end{aligned}$$

(from eqns. (105) and (45))

$$\begin{aligned}
= \mathbf{F}(V \operatorname{div} \mathbf{G} + \nabla V.\mathbf{G}) &- \mathbf{G}(V \operatorname{div} \mathbf{F} + \nabla V.\mathbf{F}) + \\
&+ V[(\mathbf{G}.\nabla)\mathbf{F} - (\mathbf{F}.\nabla)\mathbf{G}].
\end{aligned}$$

Exercises

(1) Employing the result of example (2) in §5.4, show that if ω is a constant vector, then $f(r)\omega \wedge \mathbf{r}$ is a solenoidal vector.

(2) Prove that for any pair of scalar and vector fields V and $\mathbf{F}$, $|\text{curl}\,(V\mathbf{F}) - V\,\text{curl}\,\mathbf{F}|^2 + [\text{div}\,(V\mathbf{F}) - V\,\text{div}\,\mathbf{F}]^2 = F^2|\text{grad}\,V|^2$.

(3) If the two vector fields $\mathbf{F}$ and $\mathbf{G}$ are parallel everywhere, prove that $\mathbf{F}.\text{curl}\,\mathbf{G} = \mathbf{G}.\text{curl}\,\mathbf{F}$.

5.6 DOUBLE APPLICATION OF ∇ OPERATOR

We have so far defined the operators grad, div and curl which acting respectively on scalar, vector and vector fields define new vector, scalar and vector fields. Clearly, we may now proceed to consider the effect of these three operators acting on the new fields; i.e. we shall now consider the fields arising from a double application of the ∇ operator on the original fields. It is readily shown that there are six possibilities which we shall first list, before discussing them: (1) curl grad V, (2) div curl $\mathbf{F}$, (3) grad div $\mathbf{F}$, (4) div grad $V \equiv \nabla^2 V$, (5) $\nabla^2\mathbf{F}$, (6) curl curl $\mathbf{F}$.

$$(1)\quad Curl\ Grad\ V \equiv \nabla \wedge (\nabla V)$$

This is identically zero; i.e.

$$\boxed{\text{curl grad } V \equiv 0} \tag{111}$$

for all V. This may be proved by expressing curl grad V in component form, when we obtain

$$\text{curl grad } V = \mathbf{i}\left[\frac{\partial}{\partial y}\left(\frac{\partial V}{\partial z}\right) - \frac{\partial}{\partial z}\left(\frac{\partial V}{\partial y}\right)\right] +$$
$$+ \mathbf{j}\left[\frac{\partial}{\partial z}\left(\frac{\partial V}{\partial x}\right) - \frac{\partial}{\partial x}\left(\frac{\partial V}{\partial z}\right)\right] + \mathbf{k}\left[\frac{\partial}{\partial x}\left(\frac{\partial V}{\partial y}\right) - \frac{\partial}{\partial y}\left(\frac{\partial V}{\partial x}\right)\right]$$
$$= 0,\ \text{since } \partial^2 V/\partial y\partial z = \partial^2 V/\partial z\partial y,\ \text{etc.}$$

Alternatively, employing the vector property of the ∇ operator, we have $\nabla \wedge (\nabla V) = (\nabla \wedge \nabla)V = 0$ since $\nabla \wedge \nabla = 0$.

(2) *Div Curl* $\mathbf{F} \equiv \nabla.(\nabla \wedge \mathbf{F})$

This is identically zero; i.e.

$$\text{div curl } \mathbf{F} \equiv 0 \tag{112}$$

for all $\mathbf{F}$. As in case (1), this may be proved by expressing div curl $\mathbf{F}$ in component form. Alternatively, since the triple scalar product is unaltered by cyclic permutation of its factors,

$$\nabla.(\nabla \wedge \mathbf{F}) = (\nabla \wedge \nabla).\mathbf{F} = 0.$$

(3) *Grad Div* $\mathbf{F} \equiv \nabla(\nabla.\mathbf{F})$

Expressing this in component form gives

$$\text{grad div } \mathbf{F} \equiv \mathbf{i} \left(\frac{\partial^2 F_x}{\partial x^2} + \frac{\partial^2 F_y}{\partial x \partial y} + \frac{\partial^2 F_z}{\partial x \partial z} \right) +$$

$$+ \mathbf{j} \left(\frac{\partial^2 F_x}{\partial x \partial y} + \frac{\partial^2 F_y}{\partial y^2} + \frac{\partial^2 F_z}{\partial x \partial z} \right) + \mathbf{k} \left(\frac{\partial^2 F_x}{\partial x \partial z} + \frac{\partial^2 F_y}{\partial y \partial z} + \frac{\partial^2 F_z}{\partial z^2} \right). \tag{113}$$

(4) *Div Grad* $V \equiv \nabla.(\nabla V)$

Since $\nabla.(\nabla V) = (\nabla.\nabla)V = \nabla^2 V$, div grad V is usually denoted by $\nabla^2 V$. Expressing it in component form, we obtain

$$\nabla^2 V = (\partial^2 V / \partial x^2) + (\partial^2 V / \partial y^2) + (\partial^2 V / \partial z^2).$$

$$\tag{114}$$

The operator

$$\nabla^2 \equiv (\partial^2 / \partial x^2) + (\partial^2 / \partial y^2) + (\partial^2 / \partial z^2)$$

is generally called the *Laplacian* operator.

$$(5) \quad \nabla^2 \mathbf{F}$$

We may formally consider the expression $(\nabla . \nabla)\mathbf{F}$, which is quite meaningful, although it cannot be obtained by acting with ∇ on div $\mathbf{F}$ or curl $\mathbf{F}$. It is clear from the discussion in (4) that

$$(\nabla . \nabla)\mathbf{F} = \boxed{\nabla^2 \mathbf{F} = (\partial^2 \mathbf{F}/\partial x^2) + (\partial^2 \mathbf{F}/\partial y^2) + (\partial^2 \mathbf{F}/\partial z^2),}$$

(115)

and thus in component form

$$\nabla^2 \mathbf{F} = \mathbf{i}\nabla^2 F_x + \mathbf{j}\nabla^2 F_y + \mathbf{k}\nabla^2 F_z. \tag{116}$$

$$(6) \quad \textit{Curl Curl } \mathbf{F} \equiv \nabla \wedge (\nabla \wedge \mathbf{F})$$

This satisfies the relation

$$\boxed{\text{curl curl } \mathbf{F} = \text{grad div } \mathbf{F} - \nabla^2 \mathbf{F}} \tag{117}$$

which may be proved by expanding each side in component form. Alternatively, making use of the vector properties of the ∇ operator we have

$$\text{curl curl } \mathbf{F} = \nabla \wedge (\nabla \wedge \mathbf{F}) = \nabla(\nabla . \mathbf{F}) - (\nabla . \nabla)\mathbf{F}$$

$$= \text{grad div } \mathbf{F} - \nabla^2 \mathbf{F}$$

employing the result (45).

We now discuss some consequences of the results curl grad V $\equiv$ div curl $\mathbf{F} \equiv 0$ proved above. First, we notice that of the new fields we have derived by use of the ∇ operator, grad V and grad div $\mathbf{F}$ are irrotational fields since curl grad $\equiv 0$, while curl $\mathbf{F}$ and curl curl $\mathbf{F}$ are solenoidal fields since div curl $\equiv 0$.

G

Scalar and Vector Potential Fields

At the end of §5.2 and §5.4 we broached the question of whether, given a vector field **F**, it is possible to find a scalar field V and a vector field **G** such that

$$\mathbf{F} = \text{grad } V \tag{118}$$

and/or
$$\mathbf{F} = \text{curl } \mathbf{G}. \tag{119}$$

It is now clear that a necessary condition for V to exist is that curl **F** = 0, since if V exists

$$\text{curl } \mathbf{F} = \text{curl grad } V \equiv 0.$$

We shall show later in §7.3 that this condition curl **F** = 0 is also a sufficient condition for V to exist. Similarly, a necessary condition for **G** to exist is that div **F** = 0, since if **G** exists,

$$\text{div } \mathbf{F} = \text{div curl } \mathbf{G} = 0.$$

Again, it may be shown that this condition div **F** = 0 is also a sufficient condition for **G** to exist. It follows from these results that if a vector field **F** is irrotational, then a scalar field V exists, often called a *scalar potential field*, such that **F** = grad V. If, in addition, **F** is a solenoidal vector, then since div **F** = 0, it follows that $\nabla^2 V = 0$. If, on the other hand, **F** is a solenoidal field (but not necessarily irrotational), then a vector field **G** exists, often called a *vector potential field*, such that **F** = curl **G**. It may be shown that this latter relation does *not* fix **G** uniquely, but that a unique field **G** *is* obtained if we specify div **G** independently. If in particular we let div **G** = 0, we then have

$$\text{curl } \mathbf{F} = \text{curl curl } \mathbf{G} = \text{grad div } \mathbf{G} - \nabla^2\mathbf{G} = -\nabla^2\mathbf{G}$$

from eqn. (117), since div **G** = 0. It follows then that if **F** is also irrotational, $\nabla^2\mathbf{G} = 0$. Finally, we may mention that it can be shown (although we shall not prove it here), that any vector field **F** can be written as the sum of a solenoidal and an irrotational field; i.e. for any $\mathbf{F}$ there exists a scalar field V and a vector field $\mathbf{G}$ such that

$$\mathbf{F} = \text{grad } V + \text{curl } \mathbf{G} \qquad (120)$$

with div $\mathbf{G} = 0$. Given $\mathbf{F}$, V and $\mathbf{G}$ may then be separately determined by the equations div $\mathbf{F} = \nabla^2 V$, curl $\mathbf{F} = -\nabla^2 \mathbf{G}$. We shall have occasion to employ some of these results in our work on electricity in Chapter 9.

Worked examples

(1) If $V = x^2 y^2 z^2$ and $\mathbf{F} = x^2 y\mathbf{i} + xz^3\mathbf{j} - y^2 z^2\mathbf{k}$, find
(a) $\nabla^2 V$, (b) $\nabla^2\mathbf{F}$, (c) grad div $\mathbf{F}$, (d) curl curl $\mathbf{F}$.

Ans.

(a) $\partial^2 V/\partial x^2 = 2y^2 z^2$, $\partial^2 V/\partial y^2 = 2x^2 z^2$, $\partial^2 V/\partial z^2 = 2x^2 y^2$,

and therefore

$$\nabla^2 V = (\partial^2 V/\partial x^2) + (\partial^2 V/\partial y^2) + (\partial^2 V/\partial z^2)$$
$$= 2(y^2 z^2 + x^2 z^2 + x^2 y^2).$$

(b) $\partial^2 F_x/\partial x^2 = 2y$, $\partial^2 F_x/\partial y^2 = 0 = \partial^2 F_x/\partial z^2$,

$\partial^2 F_y/\partial x^2 = 0 = \partial^2 F_y/\partial y^2$, $\partial^2 F_y/\partial z^2 = 6xz$, $\partial^2 F_z/\partial x^2 = 0$,

$\partial^2 F_z/\partial y^2 = -2z^2$, $\partial^2 F_z/\partial z^2 = -2y^2$, and

therefore

$$\nabla^2\mathbf{F} = \mathbf{i}\nabla^2 F_x + \mathbf{j}\nabla^2 F_y + \mathbf{k}\nabla^2 F_z = 2y\mathbf{i} + 6xz\mathbf{j} - 2(y^2 + z^2)\mathbf{k}.$$

(c) $\partial F_x/\partial x = 2xy$, $\partial F_y/\partial y = 0$, $\partial F_z/\partial z = -2y^2 z$, and therefore

$$\text{div } \mathbf{F} = 2(xy - y^2 z).$$

Therefore

$$\text{grad div } \mathbf{F} = 2y\mathbf{i} + 2(x - 2yz)\mathbf{j} - 2y^2\mathbf{k}.$$

(d) curl curl $\mathbf{F} = $ grad div $\mathbf{F} - \nabla^2\mathbf{F}$

$$= 2(x - 2yz - 3xz)\mathbf{j} + 2z^2\mathbf{k}.$$

(2) Prove that $\nabla^2(r^{-1}) = 0$.

Ans. $\dfrac{\partial(r^{-1})}{\partial x} = -\dfrac{1}{r^2}\dfrac{\partial r}{\partial x} = -\dfrac{x}{r^3}$, since $\dfrac{\partial r}{\partial x} = \dfrac{x}{r}$ as shown

earlier.
Therefore

$$\frac{\partial^2(r^{-1})}{\partial x^2} = -\frac{\partial}{\partial x}\frac{x}{r^3} = -\left(\frac{1}{r^3} - \frac{3x}{r^4}\frac{\partial r}{\partial x}\right) = \frac{3x^2}{r^5} - \frac{1}{r^3},$$

and similarly

$$\frac{\partial^2(r^{-1})}{\partial y^2} = \frac{3y^2}{r^5} - \frac{1}{r^3}, \quad \frac{\partial^2(r^{-1})}{\partial z^2} = \frac{3z^2}{r^5} - \frac{1}{r^3}.$$

Thus

$$\nabla^2(r^{-1}) = \frac{3x^2}{r^5} - \frac{1}{r^3} + \frac{3y^2}{r^5} - \frac{1}{r^3} + \frac{3z^2}{r^5} - \frac{1}{r^3}$$

$$= \frac{3(x^2 + y^2 + z^2)}{r^5} - \frac{3}{r^3} = 0,$$

since $r^2 = x^2 + y^2 + z^2$.

(3) Prove that for any two scalar fields U and V,

$$\text{div}\,(U\,\text{grad}\,V - V\,\text{grad}\,U) = U\nabla^2 V - V\nabla^2 U.$$

Ans. It follows from eqn. (102) that

$$\text{div}\,(U\,\text{grad}\,V) = U\,\text{div}\,\text{grad}\,V + \text{grad}\,U\,.\,\text{grad}\,V$$
$$= U\nabla^2 V + \text{grad}\,U\,.\,\text{grad}\,V.$$

Similarly,

$$\text{div}\,(V\,\text{grad}\,U) = V\nabla^2 U + \text{grad}\,U\,.\,\text{grad}\,V,\ \text{and}$$

therefore subtracting,

$$\text{div}\,(U\,\text{grad}\,V - V\,\text{grad}\,U) = U\nabla^2 V - V\nabla^2 U.$$

(4) Prove that if grad U is parallel to grad V everywhere, then curl $(U\,\text{grad}\,V) = 0$.

It follows from eqn. (104) that

curl $(U \operatorname{grad} V) = U \operatorname{curl} \operatorname{grad} V + \operatorname{grad} U \wedge \operatorname{grad} V.$

Since curl grad $V \equiv 0$, the first term on the R.H.S. is zero, and since grad U is parallel to grad V, the second term is also zero. Thus curl $(U \operatorname{grad} V) = 0$.

Exercises

(1) If $V = x^2 y^3 \exp(xz)$ and $\mathbf{F} = \sin(xyz)\mathbf{i} + (x^3 + y^3 + z^3)\mathbf{j} - x^3 y^3 z^3 \mathbf{k}$, evaluate (a) $\nabla^2 V$, (b) $\nabla^2 \mathbf{F}$, (c) grad div $\mathbf{F}$, (d) curl curl $\mathbf{F}$ and verify that curl grad $V \equiv$ div curl $\mathbf{F} \equiv 0$.

(2) Show that (a) $\nabla^2(r^n) = n(n+1)r^{n-2}$ and (b) $\nabla^2(r^n\mathbf{r}) = n(n+3)r^{n-2}\mathbf{r}$.

(3) Prove that $\nabla^2(UV) = U\nabla^2 V + 2\nabla U . \nabla V + V\nabla^2 U$.

(4) If $\mathbf{F} = \operatorname{grad} U$ and $\mathbf{G} = \operatorname{grad} V$, prove that $\mathbf{F} \wedge \mathbf{G}$ is a solenoidal vector.

(5) If $\mathbf{F} = \operatorname{curl} \mathbf{G}$, and $\mathbf{H} = \operatorname{grad} V$, show that $V\mathbf{F}$ is a solenoidal field if $\mathbf{H}$ is everywhere perpendicular to $\mathbf{F}$.

(6) By direct expansion in component form, prove that curl curl $\mathbf{F}$ $= \operatorname{grad} \operatorname{div} \mathbf{F} - \nabla^2\mathbf{F}$.

5.7 INVARIANCE PROPERTIES OF ∇

In this section, we shall discuss the question of why the specific combinations of differential coefficients given by grad, div and curl are so important. If the reader is not interested in this matter, which is rather more difficult than the other work, the section may safely be omitted without interrupting the continuity of the book.

The basic reason for the importance of the grad, div and curl operators is that they are differential operators which are *invariant* for rotation of the coordinate axes. In order to fully explain the meaning and significance of this statement, we shall divide the remainder of this section into three parts: (1) the meaning of rotational invariance, (2) the physical importance of rotational invariance, (3) the proof of the rotational invariance of grad, div, curl.

(1) *The Meaning of Rotational Invariance*

Consider the two coordinate systems (x, y, z), (x', y', z') shown in Fig. 40, which have the same origin, but which are

rotated relative to each other. Suppose now that we consider
a field J (scalar or vector) defined with respect to the (x, y, z)
axes by some given combination of differential coefficients of
another field K. For example, if J and K were scalar and vector
fields respectively, we might have

$$J = 3(\partial K_x/\partial y) - 2(\partial K_y/\partial y) + (\partial K_z/\partial x). \qquad (121)$$

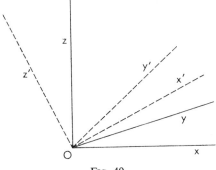

FIG. 40.

Now suppose that we wish to find the relation between J
and K when the fields are specified in terms of the (x', y', z')
axes; i.e. in the example above, we wish to find J in terms of
the differential coefficients $\partial K_x'/\partial x'$, $\partial K_x'/\partial y'$, $\partial K_y'/\partial x'$, etc.
This may readily be done (as will be shown later), knowing
the mutual orientation of the two sets of axes. However,
even without detailed calculation, it is clear that the result
obtained in this way will in general *not* be the same as that
given by replacing (x, y, z) in the original expression by
(x', y', z'); i.e. with reference to the above example,

$$J \neq 3(\partial K_x'/\partial y') - 2(\partial K_y'/\partial y') + (\partial K_z'/\partial x'). \qquad (122)$$

If, exceptionally, the result obtained with respect to the
(x', y', z') axes *is* the same as that given by replacing (x, y, z)
in the original expressions by (x', y', z'), then the operator
which acts on K to yield J is said to be invariant with respect
to rotation of axes.

(2) *The Physical Importance of Rotational Invariance*

Suppose that we consider a pair of interdependent physical quantities, vector or scalar, which are defined at each point in space and thus constitute a part of vector or scalar fields. For example, we may take a spatial distribution of electricity with change density $\rho(x, y, z)$ as one member of the pair, and the corresponding electric field $\mathbf{E}(x, y, z)$ as the other member. Let us also restrict ourselves to the case where the interdependence of the two fields is given by supposing that one is to be derived from the other by acting on the latter with a specified differential operator of the type considered in the last paragraph, and, of course, we require this relation to hold for all possible pairs of spatial distributions of the two fields. We can now proceed to prove the important result that any differential operator occurring in such a relation must be invariant for rotation of axes. To show this, we first note that with a given origin, any orientation of three mutually perpendicular cartesian coordinate axes is equally suitable for specifying the fields we are concerned with. We therefore choose two such sets of axes, rotated relatively to each other as in Fig. 40 the first set labelled (x, y, z) and the second set (x', y', z'). If a given point in space has coordinates (x, y, z), (x', y', z') respectively with respect to the two sets of axes, then a given field, J may be written $J(x, y, z)$, $J'(x', y', z')$ respectively when specified with respect to each of the two sets of axes. It is clear that $J(x, y, z) = J'(x', y', z')$ since each represents J at a definite point in space, but the functional forms J and J' will not generally be the same; i.e. $J(p, q, r) \neq J'(p, q, r)$. We are concerned with the situation when J is obtained by acting on another field K by a differential operator D, and this may be represented with respect to the first set of axes by

$$J(x, y, z) = D(x, y, z)K(x, y, z), \qquad (123)$$

and with respect to the second set by

$$J'(x', y', z') = D'(x', y', z')K'(x', y', z').$$ (124)

Here, $D'(x', y', z')$, the form that the operator takes in the second set of axes, is obtained by transforming the independent variable of the R.H.S. of eqn. (123) from x, y, z to x', y', z'. It is clear that we wish to prove that

$$D'(x', y', z') = D(x', y', z').$$

To do this, we notice that corresponding to any given pair of connected fields J_1, K_1 there always exists a second pair of connected fields J_2, K_2 which when " viewed " from the standpoint of the second coordinate system, appear to be identical with the pair J_1, K_1 as " viewed " from the standpoint of the first system. Therefore, since

$$J_1(x, y, z) = D(x, y, z)K_1(x, y, z),$$ (125)

it follows that

$$J_2'(x', y', z') = D(x', y', z')K_2'(x', y', z'),$$ (126)

since these two equations express the apparently identical relationships which are observed between J_1 and K_1 on the one hand in the first coordinate system, and J_2 and K_2 on the other hand, in the second coordinate system. Now, from eqn. (124) we have

$$J_2'(x', y', z') = D'(x', y', z')K_2'(x', y', z'),$$ (127)

and comparing this with eqn. (126) yields the required result: $D'(x', y', z') = D(x', y', z')$. We have therefore proved that if two fields corresponding to physical quantities are connected by an equation of the form (123), then the operator D must be invariant to rotation of axes. This explains why only the operators grad, div and curl are of interest in describing physical phenomena, for in the next part we shall prove their rotational invariance.

(3) *Proof of the Rotational Invariance of Grad, Div, Curl*

The rotational invariance property of grad V may be inferred indirectly from the interpretation (given in §5.2) of grad V as a vector representing both in magnitude and direction the maximum rate of change of V with respect to distance. For if we consider two coordinate systems (x, y, z) and (x', y', z'),

$$\nabla V = \mathbf{i}\frac{\partial V}{\partial x} + \mathbf{j}\frac{\partial V}{\partial y} + \mathbf{k}\frac{\partial V}{\partial z} \text{ and } \nabla' V = \mathbf{i}'\frac{\partial V}{\partial x'} + \mathbf{j}'\frac{\partial V}{\partial y'} + \mathbf{k}'\frac{\partial V}{\partial z'},$$

these being the definitions of grad V applied in each of the two systems. Now, since ∇V and $\nabla' V$ may both be shown by the above argument of §5.2 to be the vector representing the maximum rate of change of V, it follows that $\nabla V = \nabla' V$, and hence ∇V is rotationally invariant. Making use of the results of the divergence theorem considered in Chapter 7, similar proofs may be given of the rotational invariance of div and curl.

It is of interest, however, to prove these invariance results directly and this we now consider. For this purpose, we take two cartesian coordinate systems (x, y, z) and (x', y', z') as shown in Fig. 40, and shall suppose that the direction cosines with respect to the (x, y, z) axes of OX', OY', OZ' are respectively (l_1, m_1, n_1), (l_2, m_2, n_2), (l_3, m_3, n_3). This means that the direction cosines with respect to the (x', y', z') axes of OX, OY, OZ are respectively (l_1, l_2, l_3), (m_1, m_2, m_3), (n_1, n_2, n_3). If $\mathbf{i, j, k}$ and $\mathbf{i', j', k'}$ are unit vectors parallel to the (x, y, z) and (x', y', z') directions respectively, we then have, resolving $\mathbf{i', j', k'}$ in the (x, y, z) directions,

$$\left.\begin{aligned} \mathbf{i}' &= l_1\mathbf{i} + m_1\mathbf{j} + n_1\mathbf{k} \\ \mathbf{j}' &= l_2\mathbf{i} + m_2\mathbf{j} + n_2\mathbf{k} \\ \mathbf{k}' &= l_3\mathbf{i} + m_3\mathbf{j} + n_3\mathbf{k}, \end{aligned}\right\} \quad (128)$$

and resolving $\mathbf{i}, \mathbf{j}, \mathbf{k}$ in the (x', y', z') directions gives

$$\left.\begin{array}{l} \mathbf{i} = l_1\mathbf{i}' + l_2\mathbf{j}' + l_3\mathbf{k}' \\ \mathbf{j} = m_1\mathbf{i}' + m_2\mathbf{j}' + m_3\mathbf{k}' \\ \mathbf{k} = n_1\mathbf{i}' + n_2\mathbf{j}' + n_3\mathbf{k}'. \end{array}\right\} \quad (129)$$

In order to obtain the relation between the two sets of co-ordinates (x, y, z) and (x', y', z') for any point P, we express the vector $\overrightarrow{OP}$ in terms of both sets of coordinates, giving

$$\begin{aligned} x\mathbf{i} + y\mathbf{j} + z\mathbf{k} = \overrightarrow{OP} &= x'\mathbf{i}' + y'\mathbf{j}' + z'\mathbf{k}' \\ &= x'(l_1\mathbf{i} + m_1\mathbf{j} + n_1\mathbf{k}) + y'(l_2\mathbf{i} + m_2\mathbf{j} + n_2\mathbf{k}) + \\ &\quad + z'(l_3\mathbf{i} + m_3\mathbf{j} + n_3\mathbf{k}), \end{aligned}$$

substituting from eqn. (128),

$$\begin{aligned} =&\mathbf{i}(l_1x' + l_2y' + l_3z') + \mathbf{j}(m_1x' + m_2y' + m_3z') + \\ &+ \mathbf{k}(n_1x' + n_2y' + n_3z'). \end{aligned}$$

Equating the coefficients of $\mathbf{i}, \mathbf{j}, \mathbf{k}$ on the L.H.S. and R.H.S. of this equation yields

$$\left.\begin{array}{l} x = l_1x' + l_2y' + l_3z' \\ y = m_1x' + m_2y' + m_3z' \\ z = n_1x' + n_2y' + n_3z'. \end{array}\right\} \quad (130)$$

We can now proceed to show that the vector operator ∇ is invariant for rotation of axes. In the (x', y', z') system

$$\nabla' = \mathbf{i}'(\partial/\partial x') + \mathbf{j}'(\partial/\partial y') + \mathbf{k}'(\partial/\partial z'). \quad (131)$$

To change the independent variables from x', y', z' to x, y, z, we employ the result

$$\frac{\partial}{\partial x'} = \frac{\partial}{\partial x}\left(\frac{\partial x}{\partial x'}\right) + \frac{\partial}{\partial y}\left(\frac{\partial y}{\partial x'}\right) + \frac{\partial}{\partial z}\left(\frac{\partial z}{\partial x'}\right)$$

together with similar expressions for $\partial/\partial y'$ and $\partial/\partial z'$. If we then substitute $\partial x/\partial x' = l_1$, etc., from eqns. (130), we obtain

$$\partial/\partial x' = l_1(\partial/\partial x) + m_1(\partial/\partial y) + n_1(\partial/\partial z)$$

$$\partial/\partial y' = l_2(\partial/\partial x) + m_2(\partial/\partial y) + n_2(\partial/\partial z)$$

$$\partial/\partial z' = l_3(\partial/\partial x) + m_3(\partial/\partial y) + n_3(\partial/\partial z).$$

Substituting these expressions into eqn. (131) gives

$$\begin{aligned}
\nabla' = \ &\mathbf{i}'[l_1(\partial/\partial x) + m_1(\partial/\partial y) + n_1(\partial/\partial z) + \\
&+ \mathbf{j}'[l_2(\partial/\partial x) + m_2(\partial/\partial y) + n_2(\partial/\partial z)] + \\
&+ \mathbf{k}'[l_3(\partial/\partial x) + m_3(\partial/\partial y) + n_3(\partial/\partial z)] \\
= \ &(l_1\mathbf{i}' + l_2\mathbf{j}' + l_3\mathbf{k}')(\partial/\partial x) + (m_1\mathbf{i}' + m_2\mathbf{j}' + m_3\mathbf{k}')\,(\partial/\partial y) + \\
&+ (n_1\mathbf{i}' + n_2\mathbf{j}' + n_3\mathbf{k}')(\partial/\partial z) \\
= \ &\mathbf{i}(\partial/\partial x) + \mathbf{j}(\partial/\partial y) + \mathbf{k}(\partial/\partial z)
\end{aligned}$$

from eqn. (129). Thus we see that when ∇' in the (x', y', z') system is expressed in terms of (x, y, z) it yields exactly the same form with x', y', z' replaced by x, y, z. Hence the ∇ operator is invariant for rotation of the coordinate axes.

Since grad $V = \nabla V$ it follows immediately from the above result that the grad operator is invariant for rotation of axes.

Before proving the invariance of div and curl, we notice that the fact that the ∇ operator is rotationally invariant, shows that it behaves as a vector even for rotation of axes; i.e. just as the components F_x', F_y', F_z' of a vector $\mathbf{F}$ in the (x', y', z') system of axes are obtained from the corresponding components F_x, F_y, F_z in the (x, y, z) system by substituting x', y', z' for x, y, z respectively, so similarly the components $\partial/\partial x'$, $\partial/\partial y'$, $\partial/\partial z'$ of ∇ in the (x', y', z') system are obtained from the corresponding components $\partial/\partial x$, $\partial/\partial y$, $\partial/\partial z$ in the (x, y, z) system by performing a similar replacement. To prove the rotational invariance of div and curl, we first notice that for any two vectors $\mathbf{F}$ and $\mathbf{G}$, expressions of the form $\mathbf{F} . \mathbf{G}$

and $\mathbf{F} \wedge \mathbf{G}$ are rotationally invariant. This is so since the form for these expressions in the (x', y', z') system [i.e. $F_x'G_x' + F_y'G_y' + F_z'G_z'$ and $\mathbf{i}'(F_y'G_z' - F_z'G_y') + \mathbf{j}'(F_z'G_x' - F_x'G_z') + \mathbf{k}'(F_x'G_y' - F_y'G_x')$ respectively] is obtained from the corresponding expressions in the (x,y,z) system by replacing x, y, z by x', y', z' respectively. Hence since div $\mathbf{F}$ and curl $\mathbf{F}$ can be written respectively as $\nabla . F$ and $\nabla \wedge F$, and since we have shown above that ∇ behaves as a vector even for rotation of axes, it follows that div $\mathbf{F}$ and curl $\mathbf{F}$ are themselves rotationally invariant.

It is clear therefore that the importance of the grad, div and curl operators stems from the fact that, being invariant to rotation of axes, they are operators which can connect physical quantities.

6

Line, Surface and Volume
Integrals

6.1 LINE INTEGRALS

Consider a general curve C in three-dimensional space, and let the position of a variable point P on it be specified by a position vector $\mathbf{r} = \overrightarrow{OP}$ with respect to a fixed origin O as shown in Fig. 41. Then if neighbouring points P and P' have position vectors $\mathbf{r}$ and $\mathbf{r} + \delta\mathbf{r}$, it is clear that $\delta\mathbf{r} = \overrightarrow{PP'}$ is a vector representing an element of the curve in magnitude and direction. We suppose that a scalar field $V(\mathbf{r})$ and a vector field $\mathbf{F}(\mathbf{r})$ are given, and proceed to define *line integrals* of these fields along C between points A and B as follows. The portion of C lying between A and B is divided up into a large number N of elements $\delta\mathbf{r}_1, \delta\mathbf{r}_2, \ldots, \delta\mathbf{r}_p, \ldots, \delta\mathbf{r}_N$ as shown in Fig. 42, so that $\sum\limits_{p=1}^{N} \delta\mathbf{r}_p = \overrightarrow{AB}$. To define the line integral of V, we consider for any such element $\delta\mathbf{r}_p$, the vector $V(\mathbf{r})\delta\mathbf{r}_p$, where V is now taken at a point $\mathbf{r}$ lying in the elemental arc $\delta\mathbf{r}_p$. This quantity $V(\mathbf{r})\delta\mathbf{r}_p$ is then summed for all elements $\delta\mathbf{r}_p$ between A and B, and finally $\delta\mathbf{r}_p$ is allowed to tend to zero for all p as N tends to infinity. If the sum of $V(\mathbf{r})\,\delta\mathbf{r}_p$ then tends to a definite limit, this limit is called the line integral of V along the curve C between points A and B, and is denoted by $\int_A^B V(\mathbf{r}) \, d\mathbf{r}$. Thus we define

$$\int_A^B V(\mathbf{r}) \, d\mathbf{r} = \lim_{\delta r_p \to 0} \sum_{p=1}^{N} V(\mathbf{r}) \, \delta\mathbf{r}_p. \qquad (132)$$

It is clear that this integral is itself a vector.

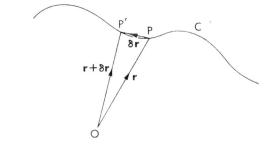

Fig. 41.

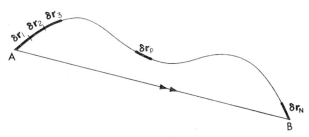

Fig. 42.

For a vector field $\mathbf{F}(\mathbf{r})$, we consider first the *scalar line integral*. This is obtained by forming the scalar product $\mathbf{F}(\mathbf{r}).\delta\mathbf{r}_p$ for each element of the curve, summing these scalars for all $\delta\mathbf{r}_p$ from A to B, and finally taking the limit as $\delta\mathbf{r}_p \to 0$. Thus the scalar line integral of $\mathbf{F}$ along C between A and B is defined by

$$\int_A^B \mathbf{F}(\mathbf{r}).\,d\mathbf{r} = \lim_{\delta r_p \to 0} \sum_{p=1}^{N} F(\mathbf{r}).\,\delta\mathbf{r}_p \qquad (133)$$

and this integral is, of course, a scalar. Finally, the *vector line integral* of **F** is defined in the same way as the scalar line integral but taking the *vector* product of **F** and $\delta\mathbf{r}_p$ instead of the *scalar* product. It is thus defined by

$$\int_A^B F(\mathbf{r}) \wedge d\mathbf{r} = \lim_{\delta r_p \to 0} \sum_{p=1}^{N} F(\mathbf{r}) \wedge \delta\mathbf{r}_p \qquad (134)$$

and is, of course, a vector. In any of these three integrals, the path of integration can be around a closed curve, in which case the points A and B coincide . This is usually denoted by writing the integral sign as $\oint$.

Evaluation of Line Integrals

In order to evaluate a given line integral it is necessary to express it in terms of variables relating to a convenient system of coordinates, for example, a suitable set of cartesian coordinates (x, y, z), and then to evaluate the resulting " ordinary " integrals. Thus, if we use cartesian coordinate (x, y, z)

$$\delta\mathbf{r} = \mathbf{i}\,\delta x + \mathbf{j}\,\delta y + \mathbf{k}\,\delta z$$

so that

(a) $V\,\delta\mathbf{r} = \mathbf{i}V\,\delta x + \mathbf{j}V\,\delta y + \mathbf{k}V\,\delta z,$

(b) $\mathbf{F}.\delta\mathbf{r} = F_x\,\delta x + F_y\,\delta y + F_z\,\delta z,$

(c) $\mathbf{F} \wedge \delta\mathbf{r} = \mathbf{i}(F_y\,\delta z - F_z\,\delta y) + \mathbf{j}(F_z\,\delta x - F_x\,\delta z) +$
$$+ \mathbf{k}(F_x\,\delta y - F_y\,\delta x).$$

Substituting these expressions in the definitions (132), (133) and (134) gives

$$\int_A^B V\,d\mathbf{r} = \mathbf{i} \int V\,dx + \mathbf{j} \int V\,dy + \mathbf{k} \int V\,dz, \qquad (135)$$

$$\int_A^B \mathbf{F}.d\mathbf{r} = \int F_x\,dx + \int F_y\,dy + \int F_z\,dz, \qquad (136)$$

$$\int_A^B \mathbf{F} \wedge d\mathbf{r} = \mathbf{i}(\int F_y\,dz - \int F_z\,dy) + \mathbf{j}(\int F_z\,dx - \int F_x\,dz) +$$
$$+ \mathbf{k}(\int F_x\,dy - \int F_y\,dx). \qquad (137)$$

It should be emphasised that the above integrands—V, F_x, F_y, F_z—are in general functions of x, y, z, and that the integrals are each to be taken with respect to x, y or z as the case may be along the given curve. This is usually done for an integral with respect to x by expressing y and z in the integrand each as functions of x from the known equation of the curve, thus reducing each integration to an " ordinary " single variable integral. Alternatively, if the parametric equations of the curve of the form $x = x(t)$, $y = y(t)$, $z = z(t)$ are known, then each integral may be immediately written as an " ordinary " integral with respect to t, since for any function $K(x, y, z)$

$$\int K(x, y, z)\, dx = \int K[x(t), y(t), z(t)](dx/dt)\, dt, \qquad (138)$$

and similarly for integrals with respect to y or z. These general methods will become clearer on working through the following examples.

Worked examples

(1) If $V(x, y, z) = x^2 y$, evaluate $\int_A^B V\, d\mathbf{r}$ along the curve with parametric equations $x = at$, $y = bt^2$, $z = ct^3$ (for constant a, b, c), where the integral is taken from $A(0, 0, 0)$ to $B(a, b, c)$.

Ans. Making use of eqns. (135), (138), and using the " dot " notation for differentiation with respect to t, we have

$$\int_A^B V\, d\mathbf{r} = \mathbf{i} \int_A^B V\dot{x}\, dt + \mathbf{j} \int_A^B V\dot{y}\, dt + \mathbf{k} \int_A^B V\dot{z}\, dt. \qquad (139)$$

Also $V = x^2 y = (at)^2(bt^2) = a^2 b t^4$, $\dot{x} = a$, $\dot{y} = 2bt$, $\dot{z} = 3ct^2$; at A, $t = 0$ and at B, $t = 1$. Therefore, substituting into eqn. (139), we obtain

$$\int_A^B V\, d\mathbf{r} = \mathbf{i} a^3 b \int_0^1 t^4\, dt + \mathbf{j} 2a^2 b^2 \int_0^1 t^5\, dt + \mathbf{k} 3a^2 bc \int_0^1 t^6\, dt$$

$$= (a^3 b/5)\mathbf{i} + (a^2 b^2/3)\mathbf{j} + (3a^2 bc/7)\mathbf{k}.$$

(2) If $\mathbf{F(r)} = 2xy\mathbf{i} + (x^2 - z^2)\mathbf{j} - 3z^2 x\mathbf{k}$, evaluate $\int_A^B \mathbf{F} . d\mathbf{r}$ between the points $A(0, 0, 0)$ and $B(2, 1, 3)$ along (a) the

curve $x = 2t$, $y = t^3$, $z = 3t^2$, (b) the straight line from $(0, 0, 0)$ to $(0, 1, 0)$, then to $(2, 1, 0)$ and finally to $(2, 1, 3)$, (c) the straight line from $(0, 0, 0)$ to $(2, 1, 3)$.

Ans. (a) Employing eqns. (136) and (138), we have

$$\int F.d\mathbf{r} = \int_A^B (F_x \dot{x} + F_y \dot{y} + F_z \dot{z})\, dt. \tag{140}$$

Here $F_x = 2xy = 2(2t)(t^3) = 4t^4$, $F_y = x^2 - z^2 = 4t^2 - 9t^4$, $F_z = -3z^2 x = -54t^5$, $\dot{x} = 2$, $\dot{y} = 3t^2$, $\dot{z} = 6t$; at A, $t = 0$ and at B, $t = 1$. Therefore we obtain

$$\int_A^B \mathbf{F}.d\mathbf{r} = \int_0^1 [2.4t^4 + 3t^2(4t^2 - 9t^4) + 6t(-54t^5)]\, dt$$

$$= \int_0^1 (20t^4 - 351t^6)\, dt = -323/7.$$

(b) Along the straight line from $A(0, 0, 0)$ to $C(0, 1, 0)$ x and z remain constant and only y alters. Therefore, of the three terms on the R.H.S. of eqn. (136) only the second one can be non-zero; i.e. $\int_A^C \mathbf{F}.d\mathbf{r} = \int_A^C F_y\, dy$. But since $F_y = x^2 - z^2$ and x, z are both zero from A to C, it follows that $\int_A^C \mathbf{F}.d\mathbf{r} = 0$. Along the straight line $C(0, 1, 0)$ to $D(2, 1, 0)$, y and z are constant and only x alters. Therefore, the second and third terms on the R.H.S. of eqn. (136) are zero, and $\int_C^D \mathbf{F}.d\mathbf{r} = \int_C^D F_x\, dx$. Now, $F_x = 2xy = 2x$ since $y = 1$ along AC, and x varies from 0 to 2. Thus

$$\int_C^D \mathbf{F}.d\mathbf{r} = 2\int_0^2 x\, dx = 4.$$

Along the straight line $D(2, 1, 0)$ to $B(2, 1, 3)$ only the final term on the R.H.S. of eqn. (136) is non-zero since x and y are constant. Thus $\int_D^B \mathbf{F}.d\mathbf{r} = \int_D^B F_z\, dz$. Now, $F_z = -3z^2 x = -6z^2$ since $x = 2$ along DB, and z varies from 0 to 3. Thus

$$\int_D^B \mathbf{F}.d\mathbf{r} = -6\int_0^3 z^2\, dz = -54.$$

Adding these terms together, we have

$$\int_A^B \mathbf{F}.d\mathbf{r} = \int_A^C \mathbf{F}.d\mathbf{r} + \int_C^D \mathbf{F}.d\mathbf{r} + \int_D^B \mathbf{F}.d\mathbf{r} = -50.$$

H

(c) The straight line from $(0, 0, 0)$ to $(2, 1, 3)$ has the parametric equations $x = 2t$, $y = t$, $z = 3t$, with t varying from 0 to 1. $F_x = 4t^2$, $F_y = -5t^2$, $F_z = -54t^3$, $\dot{x} = 2$, $\dot{y} = 1$, $\dot{z} = 3$. Thus, from eqn. (140) we obtain

$$\int \mathbf{F} . d\mathbf{r} = \int_0^1 (8t^2 - 5t^2 - 162t^3) \, dt = -79/2.$$

It should be noticed that the results of (a), (b) and (c) are all different. As we would expect, $\int_A^B \mathbf{F} . d\mathbf{r}$ depends not only on the points between which the integral is taken, but also on the path followed between the points.

(3) Find the work done in moving a particle anticlockwise once around a circle with centre $(0, 0, 3)$ and radius 5 in the $z = 3$ plane, if the force field is given by

$$\mathbf{F} = (2x + y - 2z)\mathbf{i} + (2x - 4y + z^2)\mathbf{j} + (x - 2y - z^2)\mathbf{k}.$$

Ans. It was shown in §2.3 that if a constant force $\mathbf{F}$ moved through a displacement $\mathbf{d}$, the work done is $\mathbf{F} . \mathbf{d}$. Hence if a variable force $\mathbf{F(r)}$ moves through a small displacement $\delta\mathbf{r}$, the elemental work done $= \mathbf{F} . \delta\mathbf{r}$, and so for a large displacement, the work done $= \int \mathbf{F} . d\mathbf{r}$ taken between appropriate limits. Thus in this case, the required work $W = \oint \mathbf{F} . d\mathbf{r}$, where the integral is to be taken around the given circle. Now, since all the path lies in the plane $z = 3$, $\mathbf{F}$ along the path is given by

$$\mathbf{F} = (2x + y - 6)\mathbf{i} + (2x - 4y + 9)\mathbf{j} + (x - 2y - 9)\mathbf{k}.$$

Also, the given circle has parametric equations $x = 5 \cos t$, $y = 5 \sin t$, $z = 3$, so that $\dot{x} = -5 \sin t$, $\dot{y} = 5 \cos t$, $\dot{z} = 0$ and $F_x = 10 \cos t + 5 \sin t - 6$, $F_y = 10 \cos t - 20 \sin t + 9$. The limits of t for a single passage round the circle are 0 to 2π. Hence from eqn. (140), the work W is given by

$$W = \int_0^{2\pi} [(10 \cos t + 5 \sin t - 6)(-5 \sin t) +$$
$$+ (10 \cos t - 20 \sin t + 9)(5 \cos t)] \, dt$$
$$= \int_0^{2\pi} (50 \cos^2 t - 25 \sin^2 t - 150 \sin t \cos t + 30 \sin t +$$
$$+ 45 \cos t) \, dt = 25\pi$$

since the average value of both $\cos^2 t$ and $\sin^2 t$ over a range of 2π is $\frac{1}{2}$, and the last three terms in the integrand give zero contribution to the integral.

(4) If $\mathbf{F} = xy\mathbf{i} + z^2\mathbf{j} - x\mathbf{k}$, evaluate $\oint \mathbf{F} \wedge d\mathbf{r}$ around the ellipse centre $(0, 0, 2)$ lying in the $z = 2$ plane with semi-major and minor axes, 3 and 2 in the x and y directions respectively.

Ans. The ellipse has parametric equations $x = 3 \cos t$, $y = 2 \sin t$, $z = 2$, so that $\dot{x} = -3 \sin t$, $\dot{y} = 2 \cos t$, $\dot{z} = 0$ and $F_x = 6 \sin t \cos t$, $F_y = 4$, $F_z = -3 \cos t$; the limits of t are 0 to 2π. From eqns. (137) and (139), we have

$$\oint \mathbf{F} \wedge d\mathbf{r} = \mathbf{i} \oint (F_y\dot{z} - F_z\dot{y}) \, dt + \mathbf{j} \oint (F_z\dot{x} - F_x\dot{z}) \, dt +$$
$$+ \mathbf{k} \oint (F_x\dot{y} - F_y\dot{x}) \, dt$$

$$= -\mathbf{i} \int_0^{2\pi} (-3 \cos t)(2 \cos t) \, dt + \mathbf{j} \int_0^{2\pi} (-3 \cos t) \times$$
$$\times (-3 \sin t) \, dt + \mathbf{k} \int_0^{2\pi} [(6 \sin t \cos t) \times$$
$$\times (2 \cos t) - (4)(-3 \sin t)] \, dt$$

$$= 6\mathbf{i} \int_0^{2\pi} \cos^2 t \, dt + 9\mathbf{j} \int_0^{2\pi} \cos t \sin t \, dt +$$
$$+ 12\mathbf{k} \int_0^{2\pi} (\sin t \cos^2 t + \sin t) \, dt.$$

Now,

$$\int_0^{2\pi} \cos^2 t \, dt = \pi,$$

$$\int_0^{2\pi} \cos t \sin t \, dt = 0 = \int_0^{2\pi} \sin t \, dt = \int_0^{2\pi} \sin t \cos^2 t \, dt.$$

Thus

$$\oint \mathbf{F} \wedge d\mathbf{r} = 6\pi\mathbf{i}.$$

Exercises

(1) If $V = x^2 + y^2 - z^2$, evaluate $\int V \, d\mathbf{r}$ between the points $A(0, 0, 0)$ and $B(1, 1, 2)$; (*a*) along the curve $x = t^3$, $y = t^4$, $z = 2t$, (*b*) along the straight line AB.

(2) Find the work done by the force field $\mathbf{F} = xy\mathbf{i} - z^2\mathbf{j} + xyz\mathbf{k}$ when a particle is moved from the point $A(1, 0, 2)$ to $B(3, -2, -1)$ along (*a*) the curve $x = 1 + 2t$, $y = -2t^2$, $z = 2 - 3t$, (*b*) the straight line AB, (*c*) the straight line from $(1, 0, 2)$ to $(1, 0, -1)$ then on to $(3, 0, -1)$ and finally to $(3, -2, -1)$.

(3) If $\mathbf{F} = (x^2 - z^2)\mathbf{i} - y^2\mathbf{j} + xy\mathbf{k}$, evaluate $\oint \mathbf{F}.d\mathbf{r}$ in an anticlockwise direction around the closed curve given by suitable arcs of the curves $y = x^3$, $z = 0$ and $y^2 = x$, $z = 0$.

(4) If $\mathbf{F} = x^2\mathbf{i} - xyz\mathbf{j} + yz\mathbf{k}$, evaluate $\int \mathbf{F} \wedge d\mathbf{r}$ between the points $A(0, 2, 0)$ and $B(\sqrt{2}, \sqrt{2}, 3)$ along (a) the straight line AB, (b) the circular spiral $x = 2 \sin t$, $y = 2 \cos t$, $z = (12/\pi)t$.

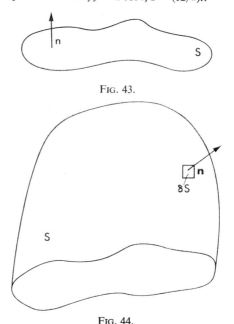

FIG. 43.

FIG. 44.

6.2 SURFACE INTEGRALS

In order to define the surface integrals which are of interest, we must first deal with the vector representation of an area. Suppose we consider a *plane* area S as shown in Fig. 43. Then the *vector area* $\mathbf{S}$ is defined as a vector perpendicular to the plane of the area and of magnitude S. The sense of the vector (to be drawn up or down) is arbitrary, but must, of course, be stated in each case. If $\mathbf{n}$ is a unit vector perpendicular to S, then clearly $\mathbf{S} = S\mathbf{n}$. In the case of a surface S

in three dimensions (shown in Fig. 44), we can ascribe a vector area $\delta\mathbf{S}$ to each elemental area δS via the definition

$$\delta\mathbf{S} = \delta S\mathbf{n}, \tag{141}$$

where $\mathbf{n}$ is a normal to the surface element δS—usually taken outwards for a closed surface.

Suppose now that a scalar field $V(\mathbf{r})$ and a vector field $\mathbf{F}(\mathbf{r})$ are given, together with a prescribed surface S (open or closed). We proceed to define surface integrals of these fields over the surface by first dividing the surface S into a large number N of elements, δS_1, δS_2, . . ., δS_p, . . ., δS_N with corresponding vector areas $\delta\mathbf{S}_1$, $\delta\mathbf{S}_2$, . . ., $\delta\mathbf{S}_p$, . . ., $\delta\mathbf{S}_N$. To define the surface integral of V, we take the product $V(\mathbf{r})\,\delta\mathbf{S}_p$ for each element δS_p and sum it for all elements δS_p of S. δS_p is then allowed to tend to zero as N tends to infinity, and if the sum of $V(\mathbf{r})\,\delta\mathbf{S}_p$ tends to a definite limit, this limit is called the *surface integral* of V over S; it is denoted by $\int_S V(\mathbf{r})\,\mathrm{d}\mathbf{S}$. Thus

$$\int_S V(\mathbf{r})\,\mathrm{d}\mathbf{S} = \lim_{\delta S_p \to 0} \sum_{p=1}^{N} V(\mathbf{r})\,\delta\mathbf{S}_p \tag{142}$$

and it is clear that this is a vector.

For a vector field $\mathbf{F}(\mathbf{r})$, we consider first the *scalar* surface integral, obtained by forming the scalar product $\mathbf{F}(\mathbf{r}) . \delta\mathbf{S}_p$ for each $\delta\mathbf{S}_p$, summing these scalars for all $\delta\mathbf{S}_p$ of the surface, and finally taking the limit as $\delta\mathbf{S}_p \to 0$. This surface integral is thus defined by

$$\int_S \mathbf{F}(\mathbf{r}) . \mathrm{d}\mathbf{S} = \lim_{\delta S_p \to 0} \sum_{p=1}^{N} \mathbf{F}(\mathbf{r}) . \delta\mathbf{S}_p \tag{143}$$

and is, of course, a scalar. Finally, the *vector* surface integral of $\mathbf{F}$ is defined in the same way as the scalar surface integral, but taking the *vector* product instead of the *scalar* product to yield

$$\int_S \mathbf{F}(\mathbf{r}) \wedge \mathrm{d}\mathbf{S} = \lim_{\delta S_p \to 0} \sum_{p=1}^{N} \mathbf{F}(\mathbf{r}) \wedge \delta\mathbf{S}_p. \tag{144}$$

Evaluation of Surface Integrals

In order to evaluate these surface integrals, we shall suppose that the cartesian equation of S is known in the form $f(x, y, z) = 0$. Then, referring to eqn. (141),

$$\mathbf{n} = \frac{\operatorname{grad} f}{\left|\operatorname{grad} f\right|}$$

since $\operatorname{grad} f$ is a vector normal to the surface $f(x, y, z) =$ constant (as shown in §5.2), and thus

$$d\mathbf{S} = dS \frac{\operatorname{grad} f}{\left|\operatorname{grad} f\right|}. \tag{145}$$

Since $\operatorname{grad} f = \mathbf{i}(\partial f/\partial x) + \mathbf{j}(\partial f/\partial y) + \mathbf{k}(\partial f/\partial z)$, it then follows from eqns. (142), (143) and (144) that

$$\int_S V \, d\mathbf{S} = \mathbf{i} \int \frac{V(\partial f/\partial x)}{\left|\operatorname{grad} f\right|} \, dS + \mathbf{j} \int \frac{V(\partial f/\partial y)}{\left|\operatorname{grad} f\right|} \, dS +$$
$$+ \mathbf{k} \int \frac{V(\partial f/\partial z)}{\left|\operatorname{grad} f\right|} \, dS, \tag{146}$$

$$\int \mathbf{F} . \, d\mathbf{S} = \int\!\!\int \left[\frac{F_x(\partial f/\partial x) + F_y(\partial f/\partial y) + F_z(\partial f/\partial z)}{\left|\operatorname{grad} f\right|} \right] dS, \tag{147}$$

$$\int \mathbf{F} \wedge d\mathbf{S} = \mathbf{i} \int\!\!\int \left[\frac{F_y(\partial f/\partial z) - F_z(\partial f/\partial y)}{\left|\operatorname{grad} f\right|} \right] dS +$$
$$+ \mathbf{j} \int\!\!\int \left[\frac{F_z(\partial f/\partial x) - F_x(\partial f/\partial z)}{\left|\operatorname{grad} f\right|} \right] dS +$$
$$+ \mathbf{k} \int \left[\frac{F_x(\partial f/\partial y) - F_y(\partial f/\partial x)}{\left|\operatorname{grad} f\right|} \right] dS. \tag{148}$$

In this way the evaluation of these surface integrals is reduced to a double integration over the surface S. The coordinates to be used for performing this double integration depend on

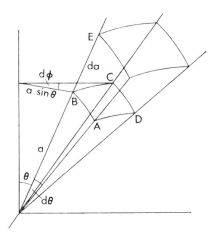

Fig. 45.

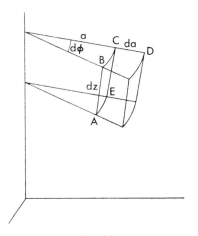

Fig. 46.

the nature of S, as will be seen presently in the worked examples. Thus, if S is the plane $z =$ constant, the x and y coordinates are usually convenient and $dS = dx\,dy$; similar remarks apply if S is the plane $x =$ constant or $y =$ constant. If S is the surface of a sphere with centre at the origin and radius a, it is convenient to use spherical polar coordinates

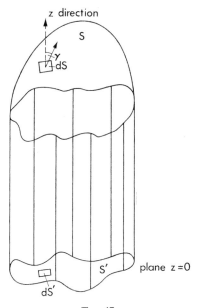

FIG. 47.

(introduced in §5.1), with (θ, φ) to specify position on the surface of the sphere. Then $x = a \sin \theta \cos \varphi$, $y = a \sin \theta \sin \varphi$, $z = a \cos \theta$, and it is readily seen from Fig. 45 that the area element

$$dS = AB \times BC = a\,d\theta \times a \sin \theta\,d\varphi = a^2 \sin \theta\,d\theta\,d\varphi.$$

If S is the curved surface of a cylinder of radius a, whose axis is the z axis, it is convenient to use cylindrical polar coordinates (introduced in §5.1) with (φ, z) to specify position on the surface of the cylinder. Then $x = a \cos \varphi$, $y = a \sin \varphi$, $z = z$, and it is readily seen from Fig. 46 that the area element

$$\mathrm{d}S = AB \times BC = \mathrm{d}z \times a\mathrm{d}\varphi = a\mathrm{d}z \, \mathrm{d}\varphi.$$

In the case of a general surface S, we may consider its projection S' on the plane $z = 0$ as shown in Fig. 47. If $\mathrm{d}S'$ is the projection of $\mathrm{d}S$ on this plane it follows that $\mathrm{d}S' = \mathrm{d}S \cos \gamma$, where γ is the angle between the normal to $\mathrm{d}S$ and the z direction. Now, since $\mathbf{n}$ is a unit vector,

$$\cos \gamma = \mathbf{n}.\mathbf{k} = (\partial f/\partial z)/\left|\operatorname{grad} f\right|,$$

whence

$$\begin{aligned}
\mathrm{d}S &= \mathrm{d}S' \left|\operatorname{grad} f\right|/(\partial f/\partial z) \\
&= \mathrm{d}x \, \mathrm{d}y[(\partial f/\partial x)^2 + (\partial f/\partial y)^2 + (\partial f/\partial z)^2]^{\frac{1}{2}}/(\partial f/\partial z),
\end{aligned}$$

$$(149)$$

taking x, y as coordinates in the $z = 0$ plane. In this way the surface integral can always be reduced to a double integration in the x–y plane.

Worked examples

(1) If $V = x^2y^2z^2$ evaluate $\int_S V \, \mathrm{d}\mathbf{S}$, where S is the curved surface of the cylinder $x^2 + y^2 = 9$ lying between the planes $z = 0$ and $z = 2$, and included in the first quadrant.

Ans. The specified surface S is shown in Fig. 48, and if $\mathbf{n}$ is a unit normal at any point

$$\mathbf{n} = \frac{\operatorname{grad}(x^2 + y^2)}{\left|\operatorname{grad}(x^2 + y^2)\right|} = \frac{2x\mathbf{i} + 2y\mathbf{j}}{(4x^2 + 4y^2)^{\frac{1}{2}}} = \frac{x\mathbf{i} + y\mathbf{j}}{3}$$

since $x^2 + y^2 = 9$ on S. Hence

$$\int_S V \, \mathrm{d}\mathbf{S} = \int_S V\mathbf{n} \, \mathrm{d}S = \frac{\mathbf{i}}{3} \int x^3y^2z^2 \, \mathrm{d}S + \frac{\mathbf{j}}{3} \int x^2y^3z^2 \, \mathrm{d}S.$$

$$(150)$$

To perform the surface integral, we use cylindrical polar coordinates $(3, \varphi, z)$ so that $x = 3 \cos \varphi$, $y = 3 \sin \varphi$ and $dS = 3 \, d\varphi \, dz$. It is clear that to cover the surface S, the limits of z are 0 to 2 and of φ are 0 to $\pi/2$. Hence from eqn. (146) we have

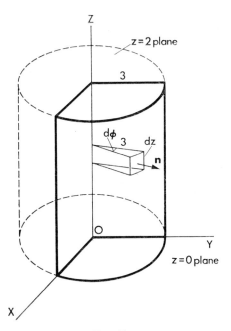

FIG. 48.

$$\int V \, dS = \mathbf{i}3^5 \int_0^2 \int_0^{\pi/2} \cos^3 \varphi \sin^2 \varphi z^2 \, d\varphi \, dz \, +$$

$$+ \, \mathbf{j}3^5 \int_0^2 \int_0^{\pi/2} \cos^2 \varphi \sin^3 \varphi z^2 \, d\varphi \, dz$$

$$= 243\mathbf{i} \int_0^{\pi/2} \cos^3 \varphi \sin^2 \varphi \, d\varphi \int_0^2 z^2 \, dz \, +$$

$$+ \, 243\mathbf{j} \int_0^{\pi/2} \cos^2 \varphi \sin^3 \varphi \, d\varphi \int_0^2 z^2 \, dz.$$

Now,

$$\int_0^{\pi/2} \cos^2 \varphi \sin^3 \varphi \, d\varphi = \int_0^{\pi/2} \cos^3 \varphi \sin^2 \varphi \, d\varphi$$

$$= \int_0^{\pi/2} (\cos^3 \varphi - \cos^5 \varphi) \, d\varphi = \frac{2}{3} - \frac{4}{5} \times \frac{2}{3} = \frac{2}{15}$$

from Wallis' formula. Also $\int_0^2 z^2 \, dz = 8/3$, whence

$$\int V \, d\mathbf{S} = 243 \times (2/15) \times (8/3)(\mathbf{i} + \mathbf{j}) = (432/5)(\mathbf{i} + \mathbf{j}).$$

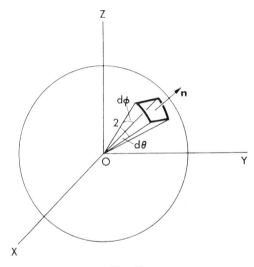

FIG. 49.

(2) If $\mathbf{F} = x\mathbf{i} + xy\mathbf{j} + xyz\mathbf{k}$, evaluate $\int \mathbf{F}.d\mathbf{S}$, where S is the complete surface of the sphere $x^2 + y^2 + z^2 = 4$.

Ans. The specified surface S is shown in Fig. 49, and if $\mathbf{n}$ is a unit normal at any point,

$$\mathbf{n} = \frac{\text{grad} \, (x^2 + y^2 + z^2)}{\left|\text{grad} \, (x^2 + y^2 + z^2)\right|} = \frac{2x\mathbf{i} + 2y\mathbf{j} + 2z\mathbf{k}}{(4x^2 + 4y^2 + 4z^2)^{\frac{1}{2}}}$$

$$= \frac{x\mathbf{i} + y\mathbf{j} + z\mathbf{k}}{2}$$

since $x^2 + y^2 + z^2 = 4$ on S. Hence

$$\int_S \mathbf{F}.\mathrm{d}\mathbf{S} = \int_S \mathbf{F}.\mathbf{n}\,\mathrm{d}S = \tfrac{1}{2} \int (x^2 + xy^2 + xyz^2)\,\mathrm{d}S. \quad (151)$$

To perform the surface integral, we use spherical polar coordinates $x = 2 \sin \theta \cos \varphi$, $y = 2 \sin \theta \sin \varphi$, $z = 2 \cos \theta$ and $\mathrm{d}S = 4 \sin \theta\,\mathrm{d}\theta\,\mathrm{d}\varphi$. To cover the surface S, the limits of θ are 0 to π and of φ are 0 to 2π. Hence from eqn. (151), we have

$$\int \mathbf{F}.\mathrm{d}\mathbf{S} = 4(\tfrac{1}{2}) \int_0^{2\pi}\int_0^{\pi}(4 \sin^2 \theta \cos^2 \varphi + 2 \sin \theta \cos \varphi \times$$
$$\times\, 4 \sin^2 \theta \sin^2 \varphi + 2 \sin \theta \cos \varphi . 2 \sin \theta \sin \varphi \times$$
$$\times\, 4 \cos^2 \theta) \sin \theta\,\mathrm{d}\theta\,\mathrm{d}\varphi$$
$$= 8 \int_0^{\pi}\sin^3 \theta\,\mathrm{d}\theta \int_0^{2\pi} \cos^2 \varphi\,\mathrm{d}\varphi +$$
$$+\, 16 \int_0^{\pi}\sin^4 \theta\,\mathrm{d}\theta \int_0^{2\pi} \cos \varphi \sin^2 \varphi\,\mathrm{d}\varphi +$$
$$+\, 32 \int_0^{\pi}\sin^2 \theta \cos^2 \theta\,\mathrm{d}\theta \int_0^{2\pi} \sin \varphi \cos \varphi\,\mathrm{d}\varphi.$$

Now,

$$\int_0^{\pi}\sin^3 \theta\,\mathrm{d}\theta = 2(2/3) = 4/3, \quad \int_0^{2\pi}\cos^2 \varphi\,\mathrm{d}\varphi = \pi$$

and

$$\int_0^{2\pi} \cos \varphi \sin^2 \varphi\,\mathrm{d}\varphi = \int_0^{2\pi} \sin \varphi \cos \varphi\,\mathrm{d}\varphi = 0.$$

Hence

$$\int \mathbf{F}.\mathrm{d}\mathbf{S} = 8(4/3)\pi + 0 + 0 = (32/3)\pi.$$

(3) If $\mathbf{F} = x\mathbf{i} - 2y\mathbf{j} - z\mathbf{k}$, evaluate $\int \mathbf{F}.\mathrm{d}\mathbf{S}$, where S is that part of the plane $x + 2y + 3z = 6$ which is located in the first octant.

Ans. The specified surface S is shown in Fig. 50, and if $\mathbf{n}$ is a unit normal at any point

$$\mathbf{n} = \frac{\mathrm{grad}\,(x + 2y + 3z)}{|\mathrm{grad}\,(x + 2y + 3z)|} = \frac{\mathbf{i} + 2\mathbf{j} + 3\mathbf{k}}{\sqrt{14}}.$$

Hence

$$\int_S \mathbf{F}.\mathrm{d}\mathbf{S} = \int_S \mathbf{F}.\mathbf{n}\,\mathrm{d}S = 14^{-\frac{1}{2}} \int_S (x - 4y - 3z)\,\mathrm{d}S.$$

To evaluate this surface integral, we consider the projection S' of S on to the $z = 0$ plane, and use coordinates (x, y) to specify position in S'. Making use of eqn. (149), with $f = x + 2y + 3z$, we obtain $dS = dx\, dy\sqrt{14}/3$, whence

$$\int_S \mathbf{F}.d\mathbf{S} = (1/3) \iint_{S'} (2x - 2y - 6)\, dx\, dy. \qquad (152)$$

since from the equation of S, $3z = 6 - x - 2y$. The area S' in the x–y plane is bounded by the positive x and y axes, and

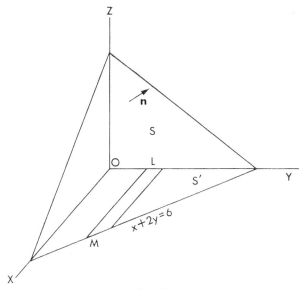

Fig. 50.

by the line $x + 2y = 6$ (since $z = 0$). Therefore to perform the double integral in eqn. (152), we keep y constant and integrate with respect to x from 0 to $6 - 2y$ (LM in Fig. 50). We then integrate with respect to y from 0 to 3 in order to completely cover S'. This yields

$$\int \mathbf{F}.d\mathbf{S} = \tfrac{1}{3} \int_0^3 \int_0^{6-2y} (2x - 2y - 6)\, dx\, dy$$
$$= \tfrac{1}{3} \int_0^3 [x^2 - 2xy - 6x]_0^{6-2y}\, dy = \tfrac{1}{3} \int_0^3 (8y^2 - 24y)\, dy$$
$$= \tfrac{1}{3}[(8/3)3^3 - (24/2)3^2] = -12.$$

(4) If $\mathbf{F} = x^2\mathbf{i} + y^2\mathbf{j} + z^2\mathbf{k}$, evaluate $\int \mathbf{F} \wedge d\mathbf{S}$, where S is that part of the plane $z = 1$ bounded by the lines $x = \pm 1$, $y = \pm 1$.

Ans. The specified surface S is shown in Fig. 51 and clearly $\mathbf{k}$ is a unit vector perpendicular to it. Thus

$$\int \mathbf{F} \wedge d\mathbf{S} = \int \mathbf{F} \wedge \mathbf{n} \, dS = \mathbf{i} \int y^2 \, dS - \mathbf{j} \int x^2 \, dS.$$

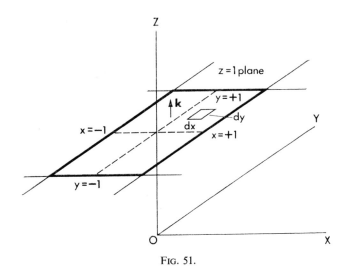

FIG. 51.

To perform the surface integrals, we use x, y as coordinates to specify position in S, and $dS = dx \, dy$. Hence

$$\int \mathbf{F} \wedge d\mathbf{S} = \mathbf{i} \int_{-1}^{+1} \int_{-1}^{+1} y^2 \, dx \, dy - \mathbf{j} \int_{-1}^{+1} \int_{-1}^{+1} x^2 \, dx \, dy$$

$$= \mathbf{i} \int_{-1}^{+1} dx \int_{-1}^{+1} y^2 \, dy - \mathbf{j} \int_{-1}^{+1} x^2 \, dx \int_{-1}^{+1} dy$$

$$= (4/3)(\mathbf{i} - \mathbf{j})$$

since

$$\int_{-1}^{+1} dx = \int_{-1}^{+1} dy = 2 \text{ and } \int_{-1}^{+1} x^2 \, dx = \int_{-1}^{+1} y^2 \, dy = 2/3.$$

Exercises

(1) If $V = x^3 + y^3 + z^3$, evaluate $\int V\,dS$ where S is the closed surface bounded by that half of the sphere $x^2 + y^2 + z^2 = 4$, for which $z > 0$, and the plane $z = 0$. [*Hint*. Evaluate $\int V\,dS$ separately for the hemisphere and the plane.]

(2) If $\mathbf{F} = x^3\mathbf{i} + y^3\mathbf{j} + z^3\mathbf{k}$, evaluate $\int \mathbf{F}.d\mathbf{S}$, where S is the curved surface of the cylinder $y^2 + z^2 = 4$ lying between the planes $x = -1$ and $x = +2$.

(3) If $\mathbf{F} = xy\mathbf{i} - 2y^2\mathbf{j} + 2xz\mathbf{k}$, evaluate $\int \mathbf{F}.d\mathbf{S}$ where S is the surface of the rectangular parallelepiped bounded by the planes $x = 0$, $x = 1$, $y = -1$, $y = +1$, $z = -1$, $z = +2$.

(4) If $\mathbf{F} = x\mathbf{i} - 2z\mathbf{j} + 3\mathbf{k}$, evaluate $\int \mathbf{F} \wedge d\mathbf{S}$, where S is that part of the plane $3x + 6y + 2z = 6$ which is situated in the first octant.

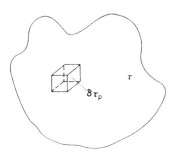

FIG. 52.

6.3   VOLUME INTEGRALS

In order to define a volume integral, consider a given volume τ divided up into N volume elements $\delta\tau_p$ $(1 \leqq p \leqq N)$ as shown in Fig. 52. It should be noted that the $\delta\tau_p$ are themselves scalars, unlike the line and area elements of §§6.1 and 6.2 which were vectors. We may then define the volume integral of a specified scalar field $V(\mathbf{r})$ by summing the products $V(\mathbf{r})\delta\tau_p$ for all volume elements $\delta\tau_p$ and taking the limit (if it exists) of this sum as $\delta\tau_p \to 0$ and $N \to \infty$. This gives the volume integral of V over τ as

$$\int_\tau V(\mathbf{r})\,d\tau = \lim_{\delta\tau_p \to 0} \sum_{p=1}^{N} V(\mathbf{r})\,\delta\tau_p, \qquad (153)$$

and this is, of course, a scalar. For a given vector field $F(r)$, the definition of the volume integral proceeds in the same way to yield

$$\int_\tau F(r)\, d\tau = \lim_{\delta\tau_p \to 0} \sum_{p=1}^{N} F(r)\, \delta\tau_p \qquad (154)$$

which is a vector. It must be realised that the notation $\int_\tau\ d\tau$ is purely an instruction to perform the corresponding volume integration by methods that we shall discuss presently. τ must not be considered as a single independent variable of which V and F are functions so that the integration with respect to τ could be done in a single step by usual integration methods.

Evaluation of Volume Integrals

The evaluation of volume integrals is performed by choosing a suitable coordinate system, for instance cartesian, spherical polars or cylindrical polars depending on the shape of the volume τ, and writing the volume integral as a triple integral in terms of these coordinates. For cartesian coordinates (x, y, z) the volume element $d\tau = dx\, dy\, dz$, for spherical polar coordinates (r, θ, φ), it is seen from Fig. 45 that

$$d\tau = AB \times BC \times BE = r\, d\theta \times r \sin\theta\, d\varphi \times dr$$

$$= r^2 \sin\theta\, dr\, d\theta\, d\varphi, \qquad (155)$$

while for cylindrical polar coordinates (ρ, φ, z) it is seen from Fig. 46 that

$$d\tau = AB \times BC \times CD = dz \times \rho\, d\varphi \times d\rho$$

$$= \rho\, d\rho\, d\varphi\, dz. \qquad (156)$$

The details of the calculation will be shown in the worked examples; meanwhile it should be noted that

$$\int_\tau F(r)\, d\tau = i \int_\tau F_x(r)\, d\tau + j \int_\tau F_y(r)\, d\tau + k \int_\tau F_z(r)\, d\tau. \qquad (157)$$

Worked examples

(1)   If $V = xy$, evaluate $\int_{\tau} V \, d\tau$ where τ is the volume bounded by the planes $x = 0, y = 0, z = 0, 3x + 2y + z = 6$.

Ans. The volume τ is shown in Fig. 53, and it is seen that the cartesian coordinates x, y, z are suitable for evaluating

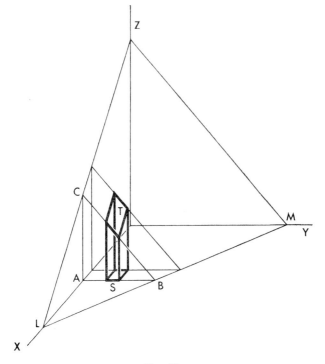

FIG. 53.

the integral. Keeping x and y constant we first integrate with respect to z, allowing it to vary from 0 to $6 - 3x - 2y$, corresponding to an integration over the column ST in Fig. 53. We then keep x constant and integrate with respect to y, allowing it to vary from 0 to $\frac{1}{2}(6 - 3x)$ corresponding to an integration of all columns over the slab ABC, since the equation of LM is $y = \frac{1}{2}(6 - 3x)$. Finally, we integrate with respect to x from 0 to 2 corresponding to an integration of all slabs over the complete volume τ. This then yields

$$\int_\tau V \, d\tau = \int_0^2 \int_0^{\frac{1}{2}(6-3x)} \int_0^{6-3x-2y} xy \, dz \, dy \, dx$$

$$= \int_0^2 \int_0^{\frac{1}{2}(6-3x)} [xyz]_0^{6-3x-2y} \, dy \, dx$$

$$= \int_0^2 \int_0^{\frac{1}{2}(6-3x)} (6xy - 3x^2y - 2xy^2) \, dy \, dx$$

$$= \int_0^2 [\tfrac{1}{2}y^2(6x - 3x^2) - \tfrac{2}{3}xy^3]_0^{\frac{1}{2}(6-3x)} \, dx$$

$$= \int_0^2 \left[(6x - 3x^2) \frac{(6 - 3x)^2}{8} - \frac{2}{3} \frac{x(6 - 3x)^3}{8} \right] dx$$

$$= 9/5.$$

(2) If $\mathbf{F} = xy\mathbf{i} + (x^2 + y^2 + z^2)\mathbf{j} - z^2\mathbf{k}$, evaluate $\int \mathbf{F} \, d\tau$ where τ is the interior of a sphere centre the origin and radius 2.

Ans. From eqn. (157) it is seen that

$$\int_\tau \mathbf{F} \, d\tau = \mathbf{i} \int_\tau xy \, d\tau + \mathbf{j} \int_\tau (x^2 + y^2 + z^2) \, d\tau - \mathbf{k} \int z^2 \, d\tau \tag{158}$$

where τ is the given sphere. Since the integration volume is spherical, it is convenient to use spherical polar coordinates (r, θ, φ), where $x = r \sin \theta \cos \varphi$, $y = r \sin \theta \sin \varphi$, $z = r \cos \theta$, $d\tau = r^2 \sin \theta \, dr \, d\theta \, d\varphi$ (see eqn. (155)). It is clear that the limits of r are 0 to 2, of θ are 0 to π, of φ are 0 to 2π. Hence from eqn. (158) we obtain

$$\int_\tau \mathbf{F} \, d\tau = \mathbf{i} \int_0^2 \int_0^\pi \int_0^{2\pi} r^2 \sin^2 \theta \sin \varphi \cos \varphi r^2 \sin \theta \, d\varphi \, d\theta \, dr +$$

$$+ \mathbf{j} \int_0^2 \int_0^\pi \int_0^{2\pi} r^2 r^2 \sin \theta \, d\varphi \, d\theta \, dr -$$

$$- \mathbf{k} \int_0^2 \int_0^\pi \int_0^{2\pi} r^2 \cos^2 \theta r^2 \sin \theta \, d\varphi \, d\theta \, dr$$

$$= \mathbf{i} \int_0^2 r^4 \, dr \int_0^\pi \sin^3 \theta \, d\theta \int_0^{2\pi} \sin \varphi \cos \varphi \, d\varphi +$$

$$+ \mathbf{j} \int_0^2 r^4 dr \int_0^\pi \sin \theta \, d\theta \int_0^{2\pi} d\varphi -$$

$$- \mathbf{k} \int_0^2 r^4 \, dr \int_0^\pi \cos^2 \theta \sin \theta \, d\theta \int_0^{2\pi} d\varphi.$$

Now, $\int_0^{2\pi}\sin\varphi\cos\varphi\,d\varphi = 0$, $\int_0^2 r^4\,dr = 32/5$, $\int_0^\pi \sin\theta\,d\theta = 2$,

$\int_0^{2\pi} d\varphi = 2\pi$, $\int_0^\pi \cos^2\theta\sin\theta\,d\theta = 2/3$, whence

$$\int \mathbf{F}\,d\tau = \mathbf{i}\times 0 + \mathbf{j}(32/5)2(2\pi) - \mathbf{k}(32/5)(2/3)(2\pi)$$

$$= (128\pi/15)(3\mathbf{j} - \mathbf{k}).$$

Exercises

(1) If $V = x^2y^2z^2$, evaluate $\int_\tau V\,d\tau$ where τ is that volume of the cylinder $x^2 + y^2 = 4$, which is intercepted between the planes $z = -1$ and $z = +2$.

(2) If $\mathbf{F} = x^3\mathbf{i} + y^3\mathbf{j} + z^3\mathbf{k}$, evaluate $\int_\tau \mathbf{F}\,d\tau$ where τ is that volume of the sphere $x^2 + y^2 + z^2 = 4$, which lies in the first octant.

(3) If $\mathbf{F} = x^2\mathbf{i} + y^2\mathbf{j} + z^2\mathbf{k}$, evaluate $\int_\tau \mathbf{F}\,d\tau$ where τ is the volume bounded by the planes $x = 0$, $y = 0$, $z = 0$, $x + 2y + 3z = 6$. [*Hint.* The x, y and z integrations may be simplified by considering carefully in which order to perform them.]

7

Theorems of Vector Integration

7.1 CONSERVATIVE VECTOR FIELDS

In §6.1 we defined the scalar line integral of a vector field, $\int_A^B \mathbf{F} \cdot d\mathbf{r}$ and saw that in general it depended not only on the extreme points A and B, but also on the path along which the integration was taken between these points. We now define:

> A *conservative* vector field $\mathbf{F}$ is one for which $\int_A^B \mathbf{F} \cdot d\mathbf{r}$ depends only on the position of A and B (for all A and B), and is independent of the path taken between them.

It is easy to show that if $\mathbf{F}$ is conservative, then for all closed paths $\oint \mathbf{F} \cdot d\mathbf{r} = 0$, since referring to Fig. 54, we see that

$$\int_{A \atop P_1}^B \mathbf{F} \cdot d\mathbf{r} = \int_{A \atop P_2}^B \mathbf{F} \cdot d\mathbf{r} = - \int_{B \atop P_2}^A \mathbf{F} \cdot d\mathbf{r}.$$

Thus

$$\int_{A \atop P_1}^B \mathbf{F} \cdot d\mathbf{r} + \int_{B \atop P_2}^A \mathbf{F} \cdot d\mathbf{r} = 0,$$

and so $\oint \mathbf{F} \cdot d\mathbf{r} = 0$. Conversely, if $\oint \mathbf{F} \cdot d\mathbf{r} = 0$ for all closed curves, then $\mathbf{F}$ is a conservative field.

We can now proceed to prove the following theorem. *A necessary and sufficient condition for $\mathbf{F}$ to be conservative is that there exists a scalar field V such that $\mathbf{F} = \text{grad } V$.* To

124

show that the condition is sufficient, assume $\mathbf{F} = \text{grad } V$. Then

$$\int_A^B \mathbf{F} . d\mathbf{r} = \int_A^B \text{grad } V . d\mathbf{r} = \int_A^B dV$$

making use of eqn. (87). But $\int_A^B dV = V_B - V_A$, where V_P is the value of V at the point P, and hence $\int_A^B \mathbf{F} . d\mathbf{r} = V_B - V_A$. Clearly the R.H.S. of this equation depends only on A and B, and is independent of the path taken between them. Hence

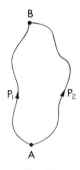

Fig. 54.

$\int_A^B \mathbf{F} . d\mathbf{r}$ is independent of the path taken between A and B, and so $\mathbf{F}$ is conservative. To show the above condition to be necessary, assume $\int_A^B \mathbf{F} . d\mathbf{r}$ is independent of the integration path between A and B. Then $\int_{\mathbf{R}'}^{\mathbf{R}} \mathbf{F}(\mathbf{r}) . d\mathbf{r} = V(\mathbf{R})$, where V depends only on $\mathbf{R}$ for given $\mathbf{R}'$. Now,

$$V(\mathbf{R} + d\mathbf{R}) - V(\mathbf{R}) = \int_{\mathbf{R}'}^{\mathbf{R} + d\mathbf{R}} \mathbf{F} . d\mathbf{r} - \int_{\mathbf{R}'}^{\mathbf{R}} \mathbf{F} . d\mathbf{r}$$

$$= \int_{\mathbf{R}}^{\mathbf{R} + d\mathbf{R}} \mathbf{F} . d\mathbf{r} = \mathbf{F} . d\mathbf{R}. \qquad (159)$$

But $V(\mathbf{R} + d\mathbf{R}) - V(\mathbf{R}) = dV(\mathbf{R}) = \text{grad } V . d\mathbf{R}$ (eqn. 87), whence by comparison with the R.H.S. of eqn. (159), we obtain $\mathbf{F} . d\mathbf{R} = \text{grad } V . d\mathbf{R}$ for *all* $d\mathbf{R}$ and so $\mathbf{F} = \text{grad } V$.

We have shown in §5.6 that if $\mathbf{F} = \text{grad } V$ then curl $\mathbf{F} = 0$, and we shall show in §7.3 that if curl $\mathbf{F} = 0$, then $\mathbf{F}$ is conservative. Therefore combining these results with that of the previous paragraph we obtain:

> Of the three statements, (1) curl $\mathbf{F} = 0$, (2) $\mathbf{F} = \text{grad } V$, (3) $\mathbf{F}$ is conservative, any one implies the other two.

It was shown in §6.1 that if $\mathbf{F}$ represents a field of force, then $\int_A^B \mathbf{F}.d\mathbf{r}$ is the work done when $\mathbf{F}$ moves a particle from A to B. It follows, therefore, that if the work is independent of the path taken between A and B, then $\mathbf{F}$ is a conservative field for which a scalar field V (the potential energy) exists such that $\mathbf{F} = \text{grad } V$; the condition for this to be so is that curl $\mathbf{F} = 0$.

Worked example

Show that the field $\mathbf{F} = (\sin y + z)\mathbf{i} + (x \cos y - z)\mathbf{j} + (x - y)\mathbf{k}$ is conservative. Obtain the field V such that $\mathbf{F} = \text{grad } V$, and hence evaluate $\int_A^B \mathbf{F}.d\mathbf{r}$ where A is $(1, 0, 2)$ and B is $(2, \pi/2, 3)$.

Ans. We prove $\mathbf{F}$ to be conservative by showing it to be irrotational; i.e. curl $\mathbf{F} = 0$. Now,

$$\text{curl } \mathbf{F} = \mathbf{i}[-1 - (-1)] + \mathbf{j}[+1 - 1] + \mathbf{k}[\cos y - \cos y] = 0,$$

and so $\mathbf{F}$ is conservative. If $\mathbf{F} = \text{grad } V$, we have $\partial V/\partial x = F_x = \sin y + z$, $\partial V/\partial y = F_y = x \cos y - z$, $\partial V/\partial z = F_z = x - y$. Hence, integrating these equations with respect to x, y and z respectively, we have

$$V = x \sin y + xz + f(y, z)$$

$$V = x \sin y - yz + g(x, z)$$

$$V = xz - yz + h(x, y)$$

where f, g, h are arbitrary functions of the given variables. Now, the forms for V given by these three equations must be identical, and it is readily seen that for this to be so

$$f(y, z) = -yz, \quad g(x, z) = xz, \quad h(x, y) = x \sin y,$$

so that $V(x, y, z) = xz - yz + x \sin y$ (omitting a possible arbitrary constant). Finally,

$$\int_A^B \mathbf{F} \cdot d\mathbf{r} = \int_A^B \text{grad } V \cdot d\mathbf{r} = V_B - V_A.$$

Now,

$$V_A = 1 \times 2 - 0 + 0 = 2$$

and

$$V_B = 2 \times 3 - 3(\pi/2) + 2 \times 1 = 8 - 3\pi/2.$$

$$\text{Thus } \int_A^B \mathbf{F} \cdot d\mathbf{r} = 6 - 3\pi/2.$$

Exercises

(1) Show that $\mathbf{F} = r^2\mathbf{r}$ is a conservative field. Illustrate this by evaluating in detail $\int_A^B \mathbf{F} \cdot d\mathbf{r}$, where A is $(0, 0, 0)$ and B is $(2, 0, 0)$, taking as integration paths (a) the straight line AB, (b) a semicircle on AB as diameter.

(2) If $\mathbf{F} = (2xy^2 + yz)\mathbf{i} + (2x^2y + xz + 2yz^2)\mathbf{j} + (2yz^2 + xy)\mathbf{k}$ is a force field, show that the work done in moving a particle from A to B is independent of the path. Hence evaluate the work done in moving a particle from $A(2, 0, 0)$ to $B(0, 2, 15\pi/2)$ along the spiral $x = 2 \cos t$, $y = 2 \sin t$, $z = 3t$. [*Hint.* Choose a simpler path between A and B for calculating the work done.]

7.2 THE DIVERGENCE THEOREM

In this section we first prove, and then discuss consequences of the divergence theorem which states that:

> Given a closed surface S enclosing a volume τ, then for any vector field $\mathbf{F}$
> $$\int_S \mathbf{F} \cdot d\mathbf{S} = \int_\tau \text{div } \mathbf{F} \, d\tau.$$ (160)

To prove this, we divide the volume τ into elemental volumes $\delta\tau$ by planes parallel to the x–y, y–z, z–x planes as shown in

Fig. 55. The elemental volumes $\delta\tau$ will be rectangular parallele-
pipeds of edge length δx, δy, δz and one of these is shown in
Fig. 56. We proceed to calculate $\int \mathbf{F}.d\mathbf{S}$ over this elemental

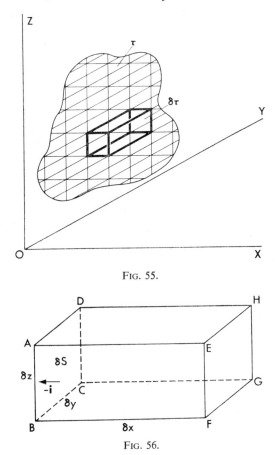

FIG. 55.

FIG. 56.

volume, and for this purpose we evaluate it first for the
opposite faces $ABCD$ and $EFGH$. Let $\delta S(= \delta y\, \delta z)$ be the
area of the face $ABCD$. Then, since we obtain $\delta\mathbf{S}$ by

multiplying δS by a unit normal pointing *outwards* for a closed surface, it follows that $\delta \mathbf{S} = -\mathbf{i}\,\delta S$, and so $\mathbf{F}.\delta \mathbf{S} = -F_x\,\delta S$. If $\delta \mathbf{S}'$ is the vector area of the face $EFGH$, then $\delta \mathbf{S}' = +\mathbf{i}\,\delta S'$ and thus $\mathbf{F}'.\delta \mathbf{S}' = +F_x\,\delta S'$, where $\mathbf{F}'$ is the value of $\mathbf{F}$ over the face $EFGH$. Therefore, summing the contributions from δS and $\delta S'$, we have

$$\mathbf{F}.\delta \mathbf{S} + \mathbf{F}'.\delta \mathbf{S}' = F_x'\,\delta S' - F_x\,\delta S = [\partial(F_x\,\delta S)/\partial x]\,\delta x \quad (161)$$

$$= (\partial F_x/\partial x)\,\delta S\,\delta x = (\partial F_x/\partial x)\,\delta x\,\delta y\,\delta z \quad (162)$$

since $\delta S = \delta y\,\delta z$ is independent of x.† Hence

$$\int_{\substack{ABCD \\ \text{and } EFGH}} \mathbf{F}.d\mathbf{S} = (\partial F_x/\partial x)\,\delta x\,\delta y\,\delta z.$$

Similarly, if we calculate $\int \mathbf{F}.d\mathbf{S}$ for the faces $AEFB$ and $DHGC$, we find

$$\int_{\substack{AEFB \\ \text{and } DHGC}} \mathbf{F}.d\mathbf{S} = (\partial F_y/\partial y)\,\delta y\,\delta x\,\delta z,$$

and for the faces $AEHD$ and $BFGC$ we find

$$\int_{\substack{AEHD \\ \text{and } BFGC}} \mathbf{F}.d\mathbf{S} = (\partial F_z/\partial z)\,\delta z\,\delta x\,\delta y.$$

Summing these three expressions then yields

$$\int_{\substack{\text{all faces of} \\ ABCDEFGH}} \mathbf{F}.d\mathbf{S} =$$

$$[(\partial F_x/\partial x) + (\partial F_y/\partial y) + (\partial F_z/\partial z)]\,\delta x\,\delta y\,\delta z = \text{div } \mathbf{F}\,\delta\tau \quad (163)$$

---

† It should be noticed that in this calculation we have taken $\partial F_x/\partial x$ to be non-zero in that F_x varies between the faces $ABCD$ and $EFGH$, but $\partial F_x/\partial y$ and $\partial F_x\partial/z$ have been taken as zero in that F_x has been assumed constant over each of the faces $ABCD$ and $EFGH$. This is permissible since the R.H.S. of eqn. (162) is non-zero only because $\partial F_x/\partial x$ has been taken as non-zero. However, it clearly remains non-zero even if $\partial F_x/\partial y$ and $\partial F_x/\partial z$ are taken to be zero. If the variation of F_x over $ABCD$ and $EFGH$ corresponding to $\partial F_x/\partial y$ and $\partial F_x/\partial z$ being non-zero had been considered, extra terms on the R.H.S. of eqn. (162) proportional to $\delta x(\delta y)^2\delta z$, etc., would have occurred, and these would vanish compared with $(\partial F_x/\partial x)\delta x\,\delta y\,\delta z$ in the limit as $\delta x,\,\delta y,\,\delta z \to 0$.

since $\delta\tau = \delta x\, \delta y\, \delta z$. Finally, we sum both sides of eqn. (163) for all elemental volumes $\delta\tau$ and then let $\delta\tau \to 0$. The R.H.S. clearly becomes $\int_\tau \operatorname{div} \mathbf{F}\, d\tau$. As far as the L.H.S. is concerned, no contribution to $\sum_{\delta\tau} \int \mathbf{F}.d\mathbf{S}$ will arise from the boundary plane between two elementary volumes (such as $ABCD$ in Fig. 56) since the plane will give equal and opposite contributions to the two elementary volumes adjoining it, and these two contributions will therefore cancel. The result of summing the L.H.S. for all $\delta\tau$ will therefore be $\int \mathbf{F}.d\mathbf{S}$ taken over the boundary surface of the whole volume, i.e. $\int_S \mathbf{F}.d\mathbf{S}$. Equating this to $\int_\tau \operatorname{div} \mathbf{F}\, d\tau$ then yields eqn. (160) as required.

Green's Theorems

Given two scalar fields U and V, Green's first theorem states:

$$\int U \operatorname{grad} V.d\mathbf{S} = \int (\operatorname{grad} U.\operatorname{grad} V + U\nabla^2 V)\, d\tau. \quad (164)$$

Green's second theorem states:

$$\int (U \operatorname{grad} V - V \operatorname{grad} U).d\mathbf{S} = \int (U\nabla^2 V - V\nabla^2 U)\, d\tau. \quad (165)$$

These theorems may be proved by use of the divergence theorem, for if $\mathbf{F} = U \operatorname{grad} V$,

$$\int U \operatorname{grad} V.d\mathbf{S} = \int \operatorname{div} (U \operatorname{grad} V)\, d\tau$$
$$= \int (\operatorname{grad} U.\operatorname{grad} V + U\nabla^2 V)\, d\tau \quad (166)$$

making use of eqn. (102); this is identical with eqn. (164). Similarly, we obtain

$$\int V \operatorname{grad} U.d\mathbf{S} = \int (\operatorname{grad} V.\operatorname{grad} U + V\nabla^2 U)\, d\tau,$$

and on subtracting this from eqn. (166), Green's second theorem is obtained.

Extensions of the Divergence Theorem

We consider now the following two theorems, related to the divergence theorem.

(1) If V is a scalar field,

$$\int_S V \, d\mathbf{S} = \int_\tau \text{grad } V \, d\tau. \tag{167}$$

(2) If $\mathbf{F}$ is a vector field,

$$\int_S d\mathbf{S} \wedge \mathbf{F} = \int_\tau \text{curl } \mathbf{F} \, d\tau. \tag{168}$$

To prove (1), we apply the divergence theorem to $V\mathbf{G}$ where $\mathbf{G}$ is an arbitrary *constant* vector. Then

$$\int V\mathbf{G} \cdot d\mathbf{S} = \int \text{div} (V\mathbf{G}) \, d\tau = \int \mathbf{G} \cdot \text{grad } V \, d\tau$$

from eqn. (102), since div $\mathbf{G} = 0$. Thus,

$$\mathbf{G} \cdot \int V \, d\mathbf{S} = \mathbf{G} \cdot \int \text{grad } V \, d\tau,$$

and since this is true for *all* constant vectors $\mathbf{G}$, eqn. (167) follows.

To prove (2), we apply the divergence theorem to $\mathbf{F} \wedge \mathbf{G}$ where $\mathbf{G}$ is an arbitrary *constant* vector. Then

$$\int (\mathbf{F} \wedge \mathbf{G}) \cdot d\mathbf{S} = \int \text{div} (\mathbf{F} \wedge \mathbf{G}) \, d\tau = \int \mathbf{G} \cdot \text{curl } \mathbf{F} \, d\tau$$

from eqn. (103), since curl $\mathbf{G} = 0$. Now

$$\int (\mathbf{F} \wedge \mathbf{G}) \cdot d\mathbf{S} = \int \mathbf{G} \cdot (d\mathbf{S} \wedge \mathbf{F})$$

from the invariance of the scalar triple product to cyclic permutations, and thus

$$\mathbf{G} \cdot \int d\mathbf{S} \wedge \mathbf{F} = \mathbf{G} \cdot \int \text{curl } \mathbf{F} \, d\tau.$$

Since this is true for *all* constant vectors $\mathbf{G}$, eqn. (168) follows.

Making use of the ∇ notation, the three eqns. (160), (167) and (168) may all be written as

$$\int_S d\mathbf{S}_* K = \int_\tau \nabla_* K \, d\tau \tag{169}$$

where K represents either a vector or scalar field, and *
represents either scalar, vector or ordinary multiplication,
as the case may be.

Integral Formulation of Grad, Div and Curl

An alternative way of expressing div **F** may be obtained by
applying the divergence theorem (160) to a small volume $\delta\tau$.
If $\delta\tau$ is small enough, div **F** may be taken as approximately
constant inside $\delta\tau$, and thus

$$\delta\tau \text{ div } \mathbf{F} \approx \int_{\delta\tau} \text{ div } \mathbf{F} \, d\tau \approx \int \mathbf{F}.d\mathbf{S},$$

where the surface integral is taken over the surface of $\delta\tau$.
Hence div $\mathbf{F} \approx \int \mathbf{F}.d\mathbf{S}/\delta\tau$ and as $\delta\tau$ becomes continually
smaller, this approximate equality becomes closer and closer
to a true equality. Thus, in the limit as $\delta\tau \to 0$

$$\text{div } \mathbf{F} = \lim_{\delta\tau \to 0} \frac{\int \mathbf{F}.d\mathbf{S}}{\delta\tau}. \tag{170}$$

By a similar approach, we see from eqns. (167) and (168) that

$$\text{grad } V = \lim_{\delta\tau \to 0} \frac{\int V \, d\mathbf{S}}{\delta\tau} \quad \text{and} \quad \text{curl } \mathbf{F} = \lim_{\delta\tau \to 0} \frac{\int d\mathbf{S} \wedge \mathbf{F}}{\delta\tau}. \tag{171}$$

These three results may all be summed up by the formula

$$\nabla_* = \lim_{\delta\tau \to 0} \frac{\int d\mathbf{S}_*}{\delta\tau}, \tag{172}$$

where each side of this operator equation acts on a vector or
scalar field as the case may be. The advantage of the forms
(170), (171) or (172) is that they make no appeal to any
(necessarily arbitrary) coordinate system in their expressions
for grad, div and curl, and it is therefore easier to glean from
them a coordinate independent interpretation of these vector

operators. In fact, it is possible to take eqns. (170) and (171) as the definitions of grad, div and curl, and to work out our previous theory directly from them. It should be noted that since these integral expressions for grad, div and curl are independent of any coordinate system, the result obtained by expressing them in a definite cartesian coordinate system must be independent of the orientation of the axes of this system. In other words, grad, div and curl when expressed in a given cartesian system are differential operators which are invariant to rotation of the coordinate axes. This result has been discussed in detail in §5.7.

Worked examples

(1) If $\mathbf{F} = x\mathbf{i} + xy\mathbf{j} + xyz\mathbf{k}$ use the divergence theorem to evaluate $\int \mathbf{F} \cdot d\mathbf{S}$, where S is the complete surface of the sphere $x^2 + y^2 + z^2 = 4$.

Ans. By the divergence theorem, $\int\limits_{S} \mathbf{F} \cdot d\mathbf{S} = \int\limits_{\tau} \operatorname{div} \mathbf{F} \, d\tau$. Here

$$\operatorname{div} \mathbf{F} = (\partial x/\partial x) + [\partial(xy)/\partial y] + [\partial(xyz)/\partial z] = 1 + x + xy,$$

and τ is the volume inside the given sphere. To evaluate the volume integral we use spherical polar coordinates (r, θ, φ) and $d\tau = r^2 \sin \theta \, dr \, d\theta \, d\varphi$. Hence

$$\int \operatorname{div} \mathbf{F} \, d\tau = \int_0^{2\pi} \int_0^\pi \int_0^2 (1 + r \sin \theta \cos \varphi +$$
$$+ r^2 \sin^2 \theta \cos \varphi \sin \varphi) r^2 \sin \theta \, dr \, d\theta \, d\varphi$$
$$= \int_0^2 r^2 \, dr \int_0^\pi \sin \theta \, d\theta \int_0^{2\pi} d\varphi +$$
$$+ \int_0^2 r^3 \, dr \int_0^\pi \sin^2 \theta \, d\theta \int_0^{2\pi} \cos \varphi \, d\varphi +$$
$$+ \int_0^2 r^4 \, dr \int_0^\pi \sin^3 \theta \, d\theta \int_0^{2\pi} \cos \varphi \sin \varphi \, d\varphi.$$

The second and third terms are each zero since

$$\int_0^{2\pi} \cos \varphi \, d\varphi = \int_0^{2\pi} \cos \varphi \sin \varphi \, d\varphi = 0.$$

For the first term, $\int_0^2 r^2 \, dr = 8/3$, $\int_0^\pi \sin\theta \, d\theta = 2$, $\int_0^{2\pi} d\varphi = 2\pi$, and so

$$\int \mathbf{F} . d\mathbf{S} = \int \operatorname{div} \mathbf{F} \, d\tau = (8/3) \times 2 \times 2\pi = 32\pi/3.$$

This is, of course, the same result as obtained by a direct surface integration in §6.2.

(2) Prove that the volume τ enclosed by a closed surface S is given by $\tau = \frac{1}{6} \int_S \operatorname{grad}(r^2) . d\mathbf{S}$.

Ans. By the divergence theorem, we have

$$\frac{1}{6} \int_S \operatorname{grad}(r^2) . d\mathbf{S} = \frac{1}{6} \int_\tau \operatorname{div} \operatorname{grad}(r^2) \, d\tau = \frac{1}{6} \int \nabla^2(r^2) \, d\tau.$$

Now,

$$\nabla^2(r^2) = [(\partial^2/\partial x^2) + (\partial^2/\partial y^2) + (\partial^2/\partial z^2)](x^2 + y^2 + z^2) = 6$$

and thus

$$\frac{1}{6} \int_S \operatorname{grad}(r^2) . d\mathbf{S} = \int_\tau d\tau = \tau.$$

(3) Prove that if the vector field $\mathbf{F(r)}$ is normal to the closed surface $G(\mathbf{r}) = \text{constant}$ at all points on the surface, then $\int_\tau \operatorname{grad} \mathbf{G} . \operatorname{curl} \mathbf{F} \, d\tau = 0$ where τ is the volume enclosed by the surface.

Ans. Since $\operatorname{grad} G$ is always normal to the surface $G(\mathbf{r}) = \text{constant}$, it follows that $\mathbf{F} \wedge \operatorname{grad} G = 0$ at all points on the surface S. Thus $\int_S (\mathbf{F} \wedge \operatorname{grad} G) . d\mathbf{S} = 0$, and so by the divergence theorem

$$\int_\tau \operatorname{div}(\mathbf{F} \wedge \operatorname{grad} G) \, d\tau = 0. \tag{173}$$

Now, from eqn. (103)

$$\operatorname{div}(\mathbf{F} \wedge \operatorname{grad} G) = \operatorname{curl} \mathbf{F} . \operatorname{grad} G - \mathbf{F} . \operatorname{curl} \operatorname{grad} G$$

$$= \operatorname{curl} \mathbf{F} . \operatorname{grad} G$$

since curl grad $G = 0$. Hence, from eqn. (173), it follows that

$$\int_\tau \text{grad } G \cdot \text{curl } \mathbf{F} \, d\tau = 0.$$

(4) A compressible fluid of density $\rho(\mathbf{r}, t)$ flows along a pipe with velocity $\mathbf{v}(\mathbf{r}, t)$, where t is the time. Prove that

$$(\partial\rho/\partial t) + \text{div}\,(\rho\mathbf{v}) = 0.$$

Ans. Consider an arbitrary volume τ of fluid surrounded by a surface S. Then the mass M of fluid in τ is given by $M = \int_\tau \rho \, d\tau$, and the time rate of change of this is

$$\partial M/\partial t = \int_\tau (\partial\rho/\partial t) \, d\tau. \qquad (174)$$

Now, the mass of fluid flowing per unit time through unit area *perpendicular* to the direction of flow is ρv and thus the mass flowing per unit time through area dS whose normal makes angle θ with the direction of flow is $\rho v \, dS \cos \theta = \rho\mathbf{v} \cdot d\mathbf{S}$. Hence the rate of loss of fluid mass from the volume τ is

$$-(\partial M/\partial t) = \int_S \rho\mathbf{v} \cdot d\mathbf{S} = \int_\tau \text{div}\,(\rho\mathbf{v}) \, d\tau.$$

Comparing this with eqn. (174), we see that

$$\int_\tau [(\partial\rho/\partial t) + \text{div}\,(\rho\mathbf{v})] \, d\tau = 0,$$

and since this is true for all volumes τ, it follows that

$$(\partial\rho/\partial t) + \text{div}\,(\rho\mathbf{v}) = 0.$$

Exercises

(1) By means of the appropriate extension of the divergence theorem, evaluate $\int_S (x^3 + y^3 + z^3) \, dS$, where S is the closed surface bounded by that half of the sphere $x^2 + y^2 + z^2 = 4$ for which $z > 0$, and the plane $z = 0$. Compare the result with that of exercise (1) §6.2.

(2) If the volume τ is surrounded by a closed surface S, prove that for a vector field $\mathbf{F}$ satisfying $\nabla^2\mathbf{F} = 0$ everywhere,
$$\int_\tau |\text{curl } \mathbf{F}|^2 \, d\tau = \int_S (\mathbf{F} \wedge \text{curl } \mathbf{F}) \cdot d\mathbf{S} + \int_\tau (\mathbf{F} \cdot \text{grad div } \mathbf{F}) \, d\tau.$$

(3) If $\mathbf{F}$ is an irrotational vector field, and V is a scalar field such that grad V is everywhere parallel to $\mathbf{F}$, prove that for any closed surface S
$$\int_S V\mathbf{F} \wedge d\mathbf{S} = 0.$$

(4) If U and V are scalar fields such that $\nabla^2 U = \nabla^2 V = 0$ everywhere' prove that for any closed surface S,
$$\int U \text{ grad } V.d\mathbf{S} = \int V \text{ grad } U.d\mathbf{S}.$$

(5) If the surface S encloses a volume τ, prove that for any *constant* vector $\mathbf{F}$,
$$\tfrac{1}{2}\int_S (\mathbf{r} \wedge \mathbf{F}) \wedge d\mathbf{S} = \mathbf{F}\tau.$$

(6) If $K(\mathbf{r})$ and $C(\mathbf{r})$ are respectively the thermal conductivity and specific heat per unit volume of a solid, prove that the temperature $T(\mathbf{r}, t)$ in the solid, at any time t, satisfies the equation
$$C(\partial T/\partial t) = K\nabla^2 T + \text{grad } K.\text{grad } T.$$

7.3 STOKES' THEOREM

In this section we first prove, and then discuss consequences of Stokes' theorem which states that:

> Given an open surface S bounded by a closed curve C, then for any vector field $\mathbf{F}$,
> $$\oint_C \mathbf{F}.d\mathbf{r} = \int_S \text{curl } \mathbf{F}.d\mathbf{S}.$$
> (175)

To prove this, we divide the surface S into elementary rectangles by means of a suitable rectangular lattice as shown in Fig. 57. We then take any such rectangular element $ABCD$, and set up a " new " coordinate system (X, Y, Z) such that the Z axis is perpendicular to the plane of $ABCD$ and the X, Y axes are parallel to AB, AD respectively as shown in Fig. 58. We proceed to calculate $\oint \mathbf{F}.d\mathbf{r}$ for $ABCD$, and for this purpose let $AB = \delta X$ and $BC = \delta Y$. Now, $\delta\mathbf{X} = \mathbf{i}\delta X$ and so $\mathbf{F}.\delta\mathbf{X} = F_X\delta X$ if $ABCD$ is traversed in an anticlockwise fashion. If $\overrightarrow{CD} = \delta\mathbf{X}'$, then $\delta\mathbf{X}' = -\mathbf{i}\,\delta X'$, and thus

$$\mathbf{F}'.\delta\mathbf{X}' = -F_X'\,\delta X',$$

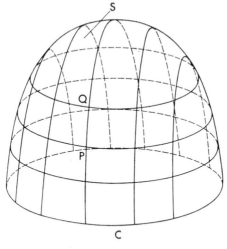

Fig. 57.

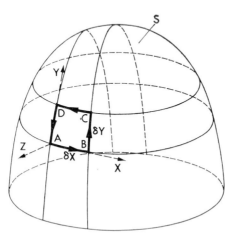

Fig. 58.

K

where $\mathbf{F'}$ is the value of $\mathbf{F}$ along CD. Hence we have

$$\mathbf{F}.\delta\mathbf{X} + \mathbf{F'}.\delta\mathbf{X'} = F_X \delta X - F_X' \delta X' = - [\partial(F_X \delta X)/\partial Y] \delta Y \tag{176}$$

$$= - (\partial F_X/\partial Y) \delta X \delta Y. \tag{177}$$

Thus we obtain

$$\int_{\substack{AB \\ \text{and } CD}} \mathbf{F}.d\mathbf{r} = - (\partial F_X/\partial Y) \delta X \delta Y.\dagger \tag{178}$$

Similarly, if $\int \mathbf{F}.d\mathbf{r}$ is calculated for the sides DA and BC of the rectangle, we obtain

$$\int_{\substack{DA \\ \text{and } BC}} \mathbf{F}.d\mathbf{r} = + (\partial F_Y/\partial X) \delta X \delta Y. \tag{179}$$

The difference in sign between eqns. (178) and (179) is due to the fact that in the latter the side BC with a larger X coordinate is traversed in the *positive* Y direction, while in the former the side CD with larger Y coordinate is traversed in the *negative* X direction. Adding the expressions (178) and (179) gives

$$\int_{ABCD} \mathbf{F}.d\mathbf{r} = [(\partial F_Y/\partial X) - (\partial F_X/\partial Y)] \delta X \delta Y$$

$$= (\text{curl } F)_Z \delta S_Z$$

where $\delta\mathbf{S} = \delta X \delta Y\mathbf{k}$ is the vector area of $ABCD$. Since $\delta S_X = \delta S_Y = 0$, we may write

$$\int_{ABCD} \mathbf{F}.d\mathbf{r} = \text{curl } \mathbf{F}.\delta\mathbf{S}. \tag{180}$$

Now, as shown in §5.7 and §7.2, curl $\mathbf{F}$ is invariant for rotation

---

† It should be noticed that in this calculation we have taken $\partial F_X/\partial Y$ to be non-zero in that F_X varies between AB and CD, but $\partial F_X/\partial X$ has been taken as zero in that F_X has been assumed constant along the side AB. This is permissible for reasons similar to those given in the footnote to eqn. (162): the R.H.S. of eqn. (177) is non-zero only by virtue of $\partial F_X/\partial Y$ being non-zero, while if $\partial F_X/\partial X$ were taken as non-zero, the change in the R.H.S. of eqn. (178) would vanish compared with $(\partial F_X/\partial Y)\delta X \delta Y$ as $\delta X, \delta Y \to 0$.

of axes, and therefore the result (180) will remain true when curl $\mathbf{F}$ and $\delta\mathbf{S}$ are expressed in terms of the usual x, y, z axes. Finally, we sum both sides of eqn. (180) for all elementary rectangles into which S is divided, and then let the area of all the rectangles tend to zero. As far as the L.H.S. is concerned no contribution to $\sum_{\text{rectangles}} \int \mathbf{F} . d\mathbf{r}$ will arise from boundary lines, such as PQ, between two rectangles, since the line will give equal and opposite contributions to the two rectangles adjoining it. The result of summing the L.H.S. will therefore be $\oint \mathbf{F} . d\mathbf{r}$ taken over the boundary curve C; i.e. $\oint_C \mathbf{F} . d\mathbf{r}$. The result of summing the R.H.S. of eqn. (180) is clearly $\int_S \text{curl } \mathbf{F} . d\mathbf{S}$, and equating this to $\oint_C \mathbf{F} . d\mathbf{r}$ gives the required result (175).

Extensions of Stokes' Theorem

We consider now the following theorems, which are related to Stokes' theorem.

(1) If V is a scalar field,

$$\oint_C V \, d\mathbf{r} = \int_S d\mathbf{S} \wedge \text{grad } V. \tag{181}$$

(2) If F is a vector field,

$$\oint_C d\mathbf{r} \wedge \mathbf{F} = \int_S (d\mathbf{S} \wedge \nabla) \wedge \mathbf{F}. \tag{182}$$

To prove (1), we apply Stokes' theorem to $V\mathbf{G}$ where $\mathbf{G}$ is an arbitrary constant vector. Then

$$\oint_C V\mathbf{G} . d\mathbf{r} = \int_S \text{curl } (V\mathbf{G}) . d\mathbf{S} = \int (\text{grad } V \wedge \mathbf{G}) . d\mathbf{S} \tag{183}$$

from eqn. (104), since curl $\mathbf{G} = 0$. Now

$$\int (\text{grad } V \wedge \mathbf{G}) . d\mathbf{S} = \int (d\mathbf{S} \wedge \text{grad } V) . \mathbf{G}$$

since the scalar triple product is invariant to cyclic permutations, and hence from eqn. (183) we obtain

$$\mathbf{G} . \oint_C V \, d\mathbf{r} = \mathbf{G} . \int_S d\mathbf{S} \wedge \text{grad } V.$$

Since this is true for *all* constant vectors **G**, eqn. (181) follows.

To prove (2), we apply Stokes' theorem to **F** ∧ **G** where **G** is an arbitrary constant vector. Then

$$\oint_C (\mathbf{F} \wedge \mathbf{G}) . d\mathbf{r} = \int_S [\nabla \wedge (\mathbf{F} \wedge \mathbf{G})] . d\mathbf{S}. \tag{184}$$

Now,

$$\oint_C (\mathbf{F} \wedge \mathbf{G}) . d\mathbf{r} = \oint_C \mathbf{G} . (d\mathbf{r} \wedge \mathbf{F}) = \mathbf{G} . \oint_C d\mathbf{r} \wedge \mathbf{F}.$$

Also, it follows from eqn. (47*a*) that

$$\int_S [\nabla \wedge (\mathbf{F} \wedge \mathbf{G})] . d\mathbf{S} = \int_S \mathbf{G} . [(d\mathbf{S} \wedge \nabla) \wedge \mathbf{F}] = \mathbf{G} . \int_S (d\mathbf{S} \wedge \nabla) \wedge \mathbf{F}$$

since **G** is a constant vector. Hence

$$\mathbf{G} . \oint_C d\mathbf{r} \wedge \mathbf{F} = \mathbf{G} . \int_S (d\mathbf{S} \wedge \nabla) \wedge \mathbf{F},$$

and since this is true for all **G**, eqn. (182) follows.

It is clear that eqns. (175), (181) and (182) may all be summed up by the formula

$$\boxed{\oint_C d\mathbf{r}_* K = \int_S (dS \wedge \nabla)_* K} \tag{185}$$

where $K = V$ or **F** and $*$ represents scalar, vector or ordinary multiplication.

Integral Formulation for Curl

An integral formulation for curl, different from the one derived in the last section, may be obtained by applying Stokes' theorem to a small *plane* surface δS surrounded by a closed curve C. If δS is small enough, curl **F** may be taken as constant inside δS, and thus

$$(\text{curl } \mathbf{F})_n \delta S \approx \int_{\delta S} \text{curl } \mathbf{F} . d\mathbf{S} \approx \oint_C \mathbf{F} . d\mathbf{r},$$

where $(\text{curl } \mathbf{F})_n$ is the resolute of curl $\mathbf{F}$ along the normal to δS. Hence $(\text{curl } \mathbf{F})_n \approx \oint \mathbf{F} \cdot d\mathbf{r}/\delta S$, and as $\delta S \rightarrow 0$ this approximate equality approaches a true equality, to give in the limit

$$(\text{curl } \mathbf{F})_n = \lim_{\delta S \to 0} \frac{\oint_C \mathbf{F} \cdot d\mathbf{r}}{\delta S}. \tag{186}$$

This result will be used in Chapter 8 to obtain the components of curl $\mathbf{F}$ in other coordinate systems.

Equivalence of Irrotational and Conservative Fields

We can now prove the result quoted in §7.1 that if curl $\mathbf{F} = 0$, then $\mathbf{F}$ is a conservative field. For, if curl $\mathbf{F} = 0$, then from Stokes' theorem

$$\oint_C \mathbf{F} \cdot d\mathbf{r} = \int_S \text{curl } \mathbf{F} \cdot d\mathbf{S} = 0$$

for all closed curves C. Hence it follows from the discussion of §7.1 that $\mathbf{F}$ is a conservative field.

Worked examples

(1)   If

$$\mathbf{F} = (2x + y - 2z)\mathbf{i} + (2x - 4y + z^2)\mathbf{j} + (x - 2y - z^2)\mathbf{k},$$

use Stokes' theorem to find $\oint_C \mathbf{F} \cdot d\mathbf{r}$, where C is the circle with centre $(0, 0, 3)$ and radius 5 in the $z = 3$ plane.

Ans. By Stokes' theorem $\oint_C \mathbf{F} \cdot d\mathbf{r} = \int_S \text{curl } \mathbf{F} \cdot d\mathbf{S}$, where S is *any* surface bounded by C. It is clear that the simplest surface to take is the interior of the circle C lying in the plane $z = 3$. Further

$$\text{curl } \mathbf{F} = \mathbf{i}(2 - 2z) + \mathbf{j}(-2 - 1) + \mathbf{k}(2 - 1)$$
$$= 2(1 - z)\mathbf{i} - 3\mathbf{j} + \mathbf{k}. \tag{187}$$

Clearly $dS = \mathbf{k}\,dS$ since $\mathbf{k}$ is a unit normal to the plane of S, and thus

$$\int_{S} \text{curl }\mathbf{F}.d\mathbf{S} = \int_{S} \text{curl }\mathbf{F}.\mathbf{k}\,dS = \int_{S} dS$$

making use of the above expression (187) for curl $\mathbf{F}$. Now

$$\int_{S} dS = S = \pi \times 5^2 = 25\pi$$

since S is a circle of radius 5, and thus $\oint_{C} \mathbf{F}.d\mathbf{r} = 25\pi$. This

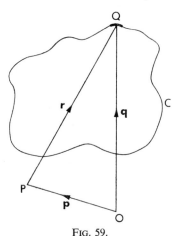

Fig. 59.

approach may be compared with that of example (3) §6.1.

(2) The value at any point P of a vector field $\mathbf{F}$ is given by $\mathbf{F} = \oint_{C} r^2\,d\mathbf{r}$, where r is the distance from P of an element $d\mathbf{r}$ of a closed curve C. Show that curl $\mathbf{F} = -4\int_{S} d\mathbf{S}$, where S is any open surface bounded by C.

Ans. The curve C is shown in Fig. 59 with $\overrightarrow{PQ} = \mathbf{r}$. If O is a fixed origin, let $\overrightarrow{OP} = \mathbf{p}$ and $\overrightarrow{OQ} = \mathbf{q}$. Then

$$\mathbf{F} = \oint_{C} r^2\,d\mathbf{r} = -\int_{S} \text{grad }(r^2)\wedge d\mathbf{S} = -2\int_{S} \mathbf{r}\wedge d\mathbf{S}$$

from eqn. (181), since grad $(r^2) = 2\mathbf{r}$. Now $\mathbf{F}$ is a function of $\mathbf{p}$, and therefore to obtain curl $\mathbf{F}$ it is necessary to perform the relevant differentiations with respect to $\mathbf{p}$ (*not* $\mathbf{r}$). Since $\mathbf{r} = \mathbf{q} - \mathbf{p}$, we have

$$\mathbf{F} = -2 \int_S (\mathbf{q} - \mathbf{p}) \wedge d\mathbf{S} = -2 \int_S \mathbf{q} \wedge d\mathbf{S} + 2\mathbf{p} \wedge \int d\mathbf{S}, \qquad (188)$$

and so

$$\text{curl } \mathbf{F} = 2 \, \text{curl}_{\mathbf{p}}(\mathbf{p} \wedge \int d\mathbf{S}), \qquad (189)$$

since the first term on the R.H.S. of eqn. (188) is independent of $\mathbf{p}$. Now, it was shown in example (1) §5.4 that for a constant vector $\boldsymbol{\omega}$, curl $(\mathbf{r} \wedge \boldsymbol{\omega}) = -2\boldsymbol{\omega}$. This result can be applied to the R.H.S. of eqn. (189) since $\int d\mathbf{S}$ is independent of $\mathbf{p}$, and hence we obtain

$$\text{curl } \mathbf{F} = 2(-2 \int d\mathbf{S}) = -4 \int_S d\mathbf{S}.$$

Exercises

(1) If $\mathbf{F} = (x^3 - y^3)\mathbf{i} + 2xy^2\mathbf{j} + xyz\mathbf{k}$, use Stokes' theorem to evaluate $\oint_C \mathbf{F}.d\mathbf{r}$, where C is the closed curve in the plane $z = 0$ bounded by the x axis, the line $x = 3$ and the curve $y = x^3$.

(2) If $\mathbf{F}$ and $\mathbf{G}$ are each irrotational vectors, prove that
$$\int_S d\mathbf{S} \wedge (\mathbf{F}.\nabla)\mathbf{G} + \int_S d\mathbf{S} \wedge (\mathbf{G}.\nabla)\mathbf{F} = \int_C \mathbf{F}.\mathbf{G} \, d\mathbf{r}.$$

(3) If $\mathbf{F} = [f(x, z) + ay]\mathbf{i} + [g(y, z) + bx]\mathbf{j} + h(x, y, z)\mathbf{k}$ for any functions f, g, h and constants a, b, show that $\oint_C \mathbf{F}.d\mathbf{r} = (b - a)S$, where the closed curve C lying in the x–y plane encloses an area S.

8

Orthogonal Curvilinear
Coordinates

8.1 VECTOR COMPONENTS IN A GENERAL
ORTHOGONAL COORDINATE SYSTEM

We introduced in §1.5 the unit vectors $\mathbf{i}$, $\mathbf{j}$, $\mathbf{k}$ parallel respectively to the x, y, z axes of a given cartesian coordinate system, and showed that any vector $\mathbf{F}$ may be expressed as $\mathbf{F} = F_x\mathbf{i} + F_y\mathbf{j} + F_z\mathbf{k}$, where $\mathbf{i}F_x$, $\mathbf{j}F_y$, $\mathbf{k}F_z$ are the components of $\mathbf{F}$ along these three directions. Now, it is clear that $\mathbf{i}$, $\mathbf{j}$, $\mathbf{k}$ is not the only triad of unit vectors which may be used for expressing a vector in terms of components, and in fact any three non-coplanar unit vectors may be used for this purpose, as was shown in §1.5. For present purposes, however, we shall restrict ourselves to the case where the unit vectors are always mutually perpendicular, but shall now consider the possibility that their orientation varies from point to point in space in some specified way. The triad of unit vectors $\mathbf{i}$, $\mathbf{j}$, $\mathbf{k}$ is clearly a special case of this where the orientation is the same everywhere; i.e. parallel to the x, y, z directions.

Spherical Polar Coordinates

In order to see how such triads of unit vectors can arise, consider the specification of a point P by means of spherical polar coordinates (r, θ, φ) as shown in Fig. 60. Then if we keep r and φ constant, and allow θ to vary, the locus of P will be the circle C shown. This circle has the vector equation $\mathbf{r} = \mathbf{r}(\theta)$ and hence it follows from the discussion in §4.1 that $\partial\mathbf{r}/\partial\theta$ at any point will be a vector along the tangent to the circle at that point. Thus $\mathbf{J} = (\partial\mathbf{r}/\partial\theta)/|\partial\mathbf{r}/\partial\theta|$ will be a unit vector along the tangent to the circle C. Similarly, if r and θ

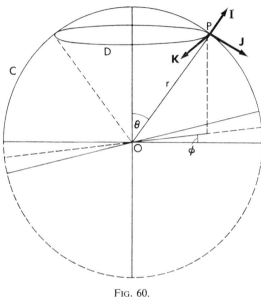

FIG. 60.

are kept constant, the locus of P as φ varies is the circle D, and thus $\mathbf{K} = (\partial\mathbf{r}/\partial\varphi)/|\partial\mathbf{r}/\partial\varphi|$ is a unit vector along the tangent to this circle. Finally, if θ and φ are kept constant, the locus of P as r varies is the straight line OP, and so $\mathbf{I} = (\partial\mathbf{r}/\partial r)/|\partial\mathbf{r}/\partial r|$ is a unit vector in the direction of OP. It is clear that $\mathbf{I}$, $\mathbf{J}$, $\mathbf{K}$ as defined above are always mutually perpendicular, and hence they constitute an example of a triad of mutually perpendicular unit vectors whose orientation varies from point to point. At a given point $P(\mathbf{r})$, their orientation is fixed, and hence a vector $\mathbf{F(r)}$ varying with position may be expressed in the form

$$\mathbf{F} = F_r\mathbf{I} + F_\theta\mathbf{J} + F_\varphi\mathbf{K} \qquad (190)$$

where $\mathbf{I}F_r$, $\mathbf{J}F_\theta$, $\mathbf{K}F_\varphi$ are the components of $\mathbf{F}$ in the $\mathbf{I}$, $\mathbf{J}$ and $\mathbf{K}$ directions. It should be noted that even if $\mathbf{F(r)}$ is independent of $\mathbf{r}$, i.e. it is a constant vector, the values of F_r, F_θ, F_φ will nevertheless vary from point to point since the orientation of $\mathbf{I}$, $\mathbf{J}$, $\mathbf{K}$ varies.

It is clear that the above approach may be employed with any coordinate system (λ, μ, v), which defines the position of a point, and it will give rise to the triad of unit vectors

$$\frac{\partial\mathbf{r}/\partial\lambda}{|\partial\mathbf{r}/\partial\lambda|}, \quad \frac{\partial\mathbf{r}/\partial\mu}{|\partial\mathbf{r}/\partial\mu|}, \quad \frac{\partial\mathbf{r}/\partial v}{|\partial\mathbf{r}/\partial v|}.$$

However, these will in general not be mutually perpendicular. If they are mutually perpendicular the coordinate system is said to be orthogonal, and several examples of such coordinate systems exist. The most important ones are the cartesian, spherical polar and cylindrical polar systems, and we shall now discuss the last of these.

Cylindrical Polar Coordinates

Let us suppose the position of a point P to be specified by cylindrical polar coordinates (ρ, φ, z), as shown in Fig. 61. Then if φ and z are kept constant, the locus of P as ρ varies is the straight line SP, and thus $\mathscr{I} = (\partial\mathbf{r}/\partial\rho)/|\partial\mathbf{r}/\partial\rho|$ is a unit vector in the direction of SP. Similarly, if ρ and z are kept constant, the locus of P as φ varies is the circle E, and thus $\mathscr{J} = (\partial\mathbf{r}/\partial\varphi)/|\partial\mathbf{r}/\partial\varphi|$ is a unit vector along the tangent to this circle. Finally, if ρ and φ are kept constant, the locus of P as z varies is the straight line TP parallel to the z axis, and thus $\mathscr{K} = (\partial\mathbf{r}/\partial z)/|\partial\mathbf{r}/\partial z|$ is a unit vector parallel to the z axis; in fact $\mathscr{K} = \mathbf{k}$. It is clear that $\mathscr{I}$, $\mathscr{J}$, $\mathscr{K}$ as defined here constitute a triad of mutually perpendicular vectors whose orientation varies from point to point. At a given point $P(\mathbf{r})$, a vector $\mathbf{F}(\mathbf{r})$ varying with position may be expressed in the form

$$\mathbf{F} = F_\rho \mathscr{I} + F_\varphi \mathscr{J} + F_z \mathscr{K}, \qquad (191)$$

where $F_\rho \mathscr{I}$, $F_\varphi \mathscr{J}$, $F_z \mathscr{K}$ are the components of $\mathbf{F}$ in the $\mathscr{I}$, $\mathscr{J}$, $\mathscr{K}$ directions.

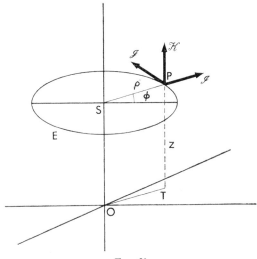

Fig. 61.

The Relation between Different Coordinate Systems

It remains to obtain the relations between the resolutes of a vector expressed in terms of two coordinate systems. The first stage in this is to find the relations between the three unit vectors of each system, and we shall now do this for the cartesian and cylindrical polar systems. For the cartesian system, we have

$$\begin{aligned}
\mathbf{r} &= x\mathbf{i} + y\mathbf{j} + z\mathbf{k} \\
&= \rho \cos \varphi \mathbf{i} + \rho \sin \varphi \mathbf{j} + z\mathbf{k}, \qquad (192)
\end{aligned}$$

expressing x, y, z in terms of ρ, φ, z. It then follows from eqn. (192) that

$$\mathscr{I} = (\partial \mathbf{r}/\partial \rho)/\left|\partial \mathbf{r}/\partial \rho\right| = (\cos \varphi \mathbf{i} + \sin \varphi \mathbf{j})/(\cos^2 \varphi + \sin^2 \varphi)^{\frac{1}{2}}$$
$$= \cos \varphi \mathbf{i} + \sin \varphi \mathbf{j}, \tag{193}$$

$$\mathscr{J} = (\partial \mathbf{r}/\partial \varphi)/\left|\partial \mathbf{r}/\partial \varphi\right|$$
$$= (-\rho \sin \varphi \mathbf{i} + \rho \cos \varphi \mathbf{j})/(\rho^2 \sin^2 \varphi + \rho^2 \cos^2 \varphi)^{\frac{1}{2}}$$
$$= - \sin \varphi \mathbf{i} + \cos \varphi \mathbf{j}, \tag{194}$$

$$\mathscr{K} = (\partial \mathbf{r}/\partial z)/\left|\partial \mathbf{r}/\partial z\right| = \mathbf{k}/1 = \mathbf{k}, \tag{195}$$

and these are the required relations between the unit vectors.

Now, for any vector field $\mathbf{F}(\mathbf{r})$ we have

$$F_x\mathbf{i} + F_y\mathbf{j} + F_z\mathbf{k} = \mathbf{F} = F_\rho\mathscr{I} + F_\varphi\mathscr{J} + F_z\mathscr{K}$$
$$= \mathbf{i}(F_\rho \cos \varphi - F_\varphi \sin \varphi) +$$
$$+ \mathbf{j}(F_\rho \sin \varphi + F_\varphi \cos \varphi) + F_z\mathbf{k}, \tag{196}$$

where in the last step we have substituted for $\mathscr{I}$, $\mathscr{J}$, $\mathscr{K}$ from eqns. (193), (194), (195). Equating then the coefficients of $\mathbf{i}$, $\mathbf{j}$, $\mathbf{k}$ on both sides of eqn. (196) yields

$$F_x = F_\rho \cos \varphi - F_\varphi \sin \varphi, \; F_y = F_\rho \sin \varphi + F_\varphi \cos \varphi, \; F_z = F_z, \tag{197}$$

which is the required relation between the resolutes of $\mathbf{F}$ in the two systems.

We now follow the same plan for the cartesian and spherical polar systems. Here

$$\mathbf{r} = x\mathbf{i} + y\mathbf{j} + z\mathbf{k}$$
$$= r \sin \theta \cos \varphi \mathbf{i} + r \sin \theta \sin \varphi \mathbf{j} + r \cos \theta \mathbf{k} \tag{198}$$

expressing x, y, z in terms of r, θ, φ. It follows then that

$$\mathbf{I} = (\partial \mathbf{r}/\partial r)/\left|\partial \mathbf{r}/\partial r\right| = \sin \theta \cos \varphi \mathbf{i} + \sin \theta \sin \varphi \mathbf{j} + \cos \theta \mathbf{k},$$
$$\mathbf{J} = (\partial \mathbf{r}/\partial \theta)/\left|\partial \mathbf{r}/\partial \theta\right| = \cos \theta \cos \varphi \mathbf{i} + \cos \theta \sin \varphi \mathbf{j} - \sin \theta \mathbf{k},$$
$$\mathbf{K} = (\partial \mathbf{r}/\partial \varphi)/\left|\partial \mathbf{r}/\partial \varphi\right| = - \sin \varphi \mathbf{i} + \cos \varphi \mathbf{j}.$$

Now, for any vector field $\mathbf{F}(\mathbf{r})$, we have

$$F_x\mathbf{i} + F_y\mathbf{j} + F_z\mathbf{k} = \mathbf{F} = F_r\mathbf{I} + F_\theta\mathbf{J} + F_\varphi\mathbf{K}$$

$$= \mathbf{i}(F_r \sin\theta \cos\varphi + F_\theta \cos\theta \cos\varphi - F_\varphi \sin\varphi) +$$

$$+ \mathbf{j}(F_r \sin\theta \sin\varphi + F_\theta \cos\theta \sin\varphi + F_\varphi \cos\varphi) +$$

$$+ \mathbf{k}(F_r \cos\theta - F_\theta \sin\theta), \tag{199}$$

substituting for $\mathbf{I}$, $\mathbf{J}$, $\mathbf{K}$ from the above relations. Hence, equating the coefficients of $\mathbf{i}$, $\mathbf{j}$, $\mathbf{k}$ on both sides of eqn. (199) gives

$$F_x = F_r \sin\theta \cos\varphi + F_\theta \cos\theta \cos\varphi - F_\varphi \sin\varphi,$$

$$F_y = F_r \sin\theta \sin\varphi + F_\theta \cos\theta \sin\varphi + F_\varphi \cos\varphi,$$

$$F_z = F_r \cos\theta - F_\theta \sin\theta. \tag{200}$$

Worked example

If a vector field $\mathbf{F}$ is given by $\mathbf{F} = r\mathbf{I} + \sec\theta\mathbf{J} + r^3 \sin\theta\mathbf{K}$, obtain F_x, F_y, F_z (in terms of x, y, z).

Ans. From eqn. (200) we have

$$F_x = r \sin\theta \cos\varphi + \cos\varphi - r^3 \sin\theta \sin\varphi$$

$$= x + [x/(x^2 + y^2)^{\frac{1}{2}}] - y(x^2 + y^2 + z^2),$$

$$F_y = r \sin\theta \sin\varphi + \sin\varphi + r^3 \sin\theta \cos\varphi$$

$$= y + [y/(x^2 + y^2)^{\frac{1}{2}}] + x(x^2 + y^2 + z^2),$$

$$F_z = r \cos\theta - \tan\theta = z - [(x^2 + y^2)^{\frac{1}{2}}/z].$$

Exercise

If a vector field $\mathbf{F}$ is given by $\mathbf{F} = 2z\mathbf{i} + 2x\mathbf{j} - 3y^2\mathbf{k}$, obtain F_ρ, F_φ, F_z (in terms of ρ, φ, z).

8.2 DIFFERENTIAL OPERATORS FOR ORTHOGONAL COORDINATES

In the last section we introduced the use of orthogonal coordinates, and in particular polar coordinates, for specifying the components of a vector field. We now wish to obtain explicit expressions for the grad, div and curl operators in these systems. The reason for doing this is that in many physical problems the system has either spherical or cylindrical symmetry, and calculations can then be done most easily by working throughout in the appropriate coordinate system.

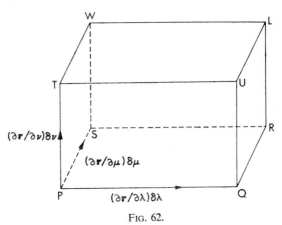

Fig. 62.

To obtain the expressions for these differential operators, we shall use a general orthogonal coordinate system (λ, μ, v) which covers, of course, both cylindrical and spherical coordinates. Let us first consider the eight points $P(\lambda, \mu, v)$, $Q(\lambda + \delta\lambda, \mu, v)$, $R(\lambda + \delta\lambda, \mu + \delta\mu, v)$, $S(\lambda, \mu + \delta\mu, v)$, $T(\lambda, \mu, v + \delta v)$, $U(\lambda + \delta\lambda, \mu, v + \delta v)$, $L(\lambda + \delta\lambda, \mu + \delta\mu, v + \delta v)$, $W(\lambda, \mu + \delta\mu, v + \delta v)$ shown in Fig. 62. Then if $\overrightarrow{OP} = \mathbf{r}$,

$$\overrightarrow{PQ} = (\partial\mathbf{r}/\partial\lambda)\delta\lambda, \quad \overrightarrow{PS} = (\partial\mathbf{r}/\partial\mu)\delta\mu, \quad \overrightarrow{PT} = (\partial\mathbf{r}/\partial v)\delta v. \quad (201)$$

Now, since the coordinate system is an orthogonal one, $(\partial \mathbf{r}/\partial \lambda)$, $(\partial \mathbf{r}/\partial \mu)$, $(\partial \mathbf{r}/\partial v)$ are mutually perpendicular, and hence $PQRSTULW$ is a rectangular parallelepiped.
Further,

$$
\begin{aligned}
PQ &= \left|\partial \mathbf{r}/\partial \lambda\right|\delta \lambda = h_\lambda \delta \lambda, \text{ where } h_\lambda = \left|\partial \mathbf{r}/\partial \lambda\right|; \\
PS &= \left|\partial \mathbf{r}/\partial \mu\right|\delta \mu = h_\mu \delta \mu, \text{ where } h_\mu = \left|\partial \mathbf{r}/\partial \mu\right|; \\
PT &= \left|\partial \mathbf{r}/\partial v\right|\delta v = h_v \delta v, \text{ where } h_v = \left|\partial \mathbf{r}/\partial v\right|.
\end{aligned}
\quad (202)
$$

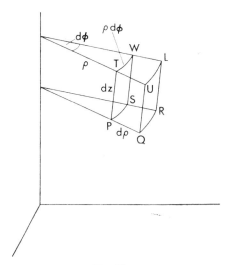

Fig. 63.

Applying these results to the case of cylindrical polars shown in Fig. 63, we readily see from eqns. (193), (194), (195), (202) that if $(\lambda, \mu, v) \equiv (\rho, \varphi, z)$, then

$$
h_\rho = 1, \quad h_\varphi = \rho, \quad h_z = 1. \quad (203)
$$

Similarly, in the case of spherical polars, shown in Fig. 64, we see from eqn. (202) that if $(\lambda, \mu, v) \equiv (r, \theta, \varphi)$, then

$$
h_r = 1, \quad h_\theta = r, \quad h_\varphi = r \sin \theta. \quad (204)
$$

The Grad Operator

Consider the change in value dV of a scalar field $V(\mathbf{r})$ in going from P to L. We may apply eqn. (87), and if $\overrightarrow{PL} = d\mathbf{r}$, we have $d\mathbf{r} = (\partial\mathbf{r}/\partial\lambda)\delta\lambda + (\partial\mathbf{r}/\partial\mu)\delta\mu + (\partial\mathbf{r}/\partial v)\delta v$ from eqn. (201), since $\overrightarrow{PL} = \overrightarrow{PQ} + \overrightarrow{PS} + \overrightarrow{PT}$. Hence

$$dV = (\text{grad } V)_\lambda |\partial\mathbf{r}/\partial\lambda|\delta\lambda + (\text{grad } V)_\mu |\partial\mathbf{r}/\partial\mu|\delta\mu +$$
$$+ (\text{grad } V)_v |\partial\mathbf{r}/\partial v|\delta v$$

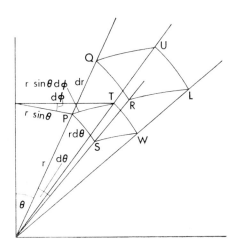

Fɪɢ. 64.

since $\partial\mathbf{r}/\partial\lambda$, $\partial\mathbf{r}/\partial\mu$, $\partial\mathbf{r}/\partial v$ are mutually perpendicular. Thus

$$dV = h_\lambda(\text{grad } V)_\lambda\delta\lambda + h_\mu(\text{grad } V)_\mu\delta\mu + h_v(\text{grad } V)_v\delta v$$

from eqn. (202), and so

$$(\text{grad } V)_\lambda = h_\lambda^{-1}\,(\partial V/\partial\lambda), \quad (\text{grad } V)_\mu = h_\mu^{-1}\,(\partial V/\partial\mu),$$
$$(\text{grad } V)_v = h_v^{-1}\,(\partial V/\partial v). \quad (205)$$

Hence

$$\text{grad } V = h_\lambda^{-1} (\partial V/\partial\lambda)\boldsymbol{\alpha} + h_\mu^{-1} (\partial V/\partial\mu)\boldsymbol{\beta} +$$
$$+ h_\nu^{-1}(\partial V/\partial\nu)\boldsymbol{\gamma} \qquad (206)$$

where $\boldsymbol{\alpha}$, $\boldsymbol{\beta}$, $\boldsymbol{\gamma}$ are unit vectors along the direction of $(\partial\mathbf{r}/\partial\lambda)$, $(\partial\mathbf{r}/\partial\mu)$, $(\partial\mathbf{r}/\partial\nu)$ respectively.

Applying eqn. (206) to cylindrical polars gives

$$\text{grad } V = (\partial V/\partial\rho)\mathscr{I} + \rho^{-1}(\partial V/\partial\varphi)\mathscr{J} + (\partial V/\partial z)\mathscr{K},$$

while for spherical polars we obtain

$$\text{grad } V = (\partial V/\partial r)\mathbf{I} + r^{-1}(\partial V/\partial\theta)\mathbf{J} + r^{-1} \text{ cosec } \theta(\partial V/\partial\varphi)\mathbf{K}.$$

The Div Operator

We obtain the expression for div $\mathbf{F}$ in terms of (λ, μ, ν) by applying eqn. (170) to the rectangular parallelepiped *PQRSTULW*. It is clear that since this parallelepiped is rectangular

$$\delta\tau = PQ \times PS \times PT = h_\lambda h_\mu h_\nu \, \delta\lambda \, \delta\mu \, \delta\nu \qquad (207)$$

from eqn. (202). To calculate $\int \mathbf{F}.\mathbf{dS}$ for the surface of *PQRSTULW*, we use a similar approach to that employed in proving the first part of the divergence theorem in §7.2, and begin by evaluating $\int \mathbf{F}.\mathbf{dS}$ for the faces *PSWT* and *QRLU*. If the area of the face *PSWT* $= \delta S$, then it follows in complete analogy with the derivation of eqn. (161) that

$$\int_{\substack{PSWT \\ \text{and } QRLU}} \mathbf{F}.\mathbf{dS} = [\partial(F_\lambda\delta S)/\partial\lambda] \, \delta\lambda$$

$$= [\partial(h_\mu h_\nu F_\lambda)/\partial\lambda] \, \delta\lambda \, \delta\mu \, \delta\nu \qquad (208)$$

since $\delta S = PS \times PT = h_\mu h_\nu \, \delta\mu \, \delta\nu$, and $\delta\mu$, $\delta\nu$ are independent

L

of λ†. Similarly, if we calculate $\int \mathbf{F} \cdot d\mathbf{S}$ for the pair of faces $PQUT$, $SRLW$ and for the pair of faces $PQRS$, $TULW$, we obtain

$$\int_{\substack{PQUT \\ \text{and } SRLW}} \mathbf{F} \cdot d\mathbf{S} = [\partial(h_\lambda h_\nu F_\mu)/\partial\mu] \, \delta\lambda \, \delta\mu \, \delta\nu, \tag{209}$$

$$\int_{\substack{PQRS \\ \text{and } TULW}} \mathbf{F} \cdot d\mathbf{S} = [\partial(h_\lambda h_\mu F_\nu)/\partial\nu] \, \delta\lambda \, \delta\mu \, \delta\nu. \tag{210}$$

Summing the three expressions (208), (209) and (210) gives finally

$$\int_{\substack{\text{all faces of} \\ PQRSTULW}} \mathbf{F} \cdot d\mathbf{S} = \{[\partial(h_\mu h_\nu F_\lambda)/\partial\lambda] + [\partial(h_\lambda h_\nu F_\mu)/\partial\mu] +$$

$$+ [\partial(h_\lambda h_\mu F_\nu/\partial\nu]\} \, \delta\lambda \, \delta\mu \, \delta\nu.$$

Dividing this by expression (207) for $\delta\tau$ and going to the limit as $\delta\lambda$, $\delta\mu$, $\delta\nu \to 0$ then yields

$$\text{div } \mathbf{F} = \lim_{\delta\tau \to 0} \frac{\int \mathbf{F} \cdot d\mathbf{S}}{\delta\tau} = \frac{1}{h_\lambda h_\mu h_\nu} \left[\frac{\partial(h_\mu h_\nu F_\lambda)}{\partial\lambda} + \frac{\partial(h_\lambda h_\nu F_\mu)}{\partial\mu} + \right.$$

$$\left. + \frac{\partial(h_\lambda h_\mu F_\nu)}{\partial\nu} \right]. \tag{211}$$

Combining this result with eqn. (206) gives

$$\nabla^2 V = \text{div grad } V = \frac{1}{h_\lambda h_\mu h_\nu} \left[\frac{\partial}{\partial\lambda} \left(\frac{h_\mu h_\nu}{h_\lambda} \frac{\partial V}{\partial\lambda} \right) + \right.$$

$$\left. + \frac{\partial}{\partial\mu} \left(\frac{h_\lambda h_\nu}{h_\mu} \frac{\partial V}{\partial\mu} \right) + \frac{\partial}{\partial\nu} \left(\frac{h_\lambda h_\mu}{h_\nu} \frac{\partial V}{\partial\nu} \right) \right]. \tag{212}$$

---

† It should be noticed that on the R.H.S. of eqn. (208) we have taken into account the fact that the area of $PSWT$ is not exactly equal to that of $QRLU$ in that there is a term arising from $\partial(\delta S)/\partial\lambda \propto \partial(h_\mu h_\nu)/\partial\lambda$ which is not taken to be zero. It is necessary to consider this term since it gives a contribution to the R.H.S. of eqn. (208) of the same order of magnitude as that arising from $\partial F_\lambda/\partial\lambda$. However, as far as the volume $\delta\tau$ is concerned, we may take $PQRSTULW$ to be a perfect parallelepiped, since the actual deviation from this shape yields an increment in volume whose ratio to the true volume tends to zero as $\delta\lambda$, $\delta\mu$, $\delta\nu \to 0$.

The results (211) and (212) may be immediately applied to the case of cylindrical and spherical polar coordinates, when we obtain from eqns. (211) and (212) that for cylindrical polars,

$$\operatorname{div} \mathbf{F} = \rho^{-1}[\partial(\rho F_\rho)/\partial\rho] + \rho^{-1}(\partial F_\varphi/\partial\varphi) + \partial F_z/\partial z, \quad (213)$$

$$\nabla^2 V = \frac{1}{\rho}\frac{\partial}{\partial\rho}\left(\rho\frac{\partial V}{\partial\rho}\right) + \frac{1}{\rho^2}\frac{\partial^2 V}{\partial\varphi^2} + \frac{\partial^2 V}{\partial z^2}, \quad (214)$$

and for spherical polars

$$\operatorname{div} \mathbf{F} = \frac{1}{r^2}\frac{\partial}{\partial r}(r^2 F_r) + \frac{1}{r\sin\theta}\frac{\partial}{\partial\theta}(\sin\theta\, F_\theta) + \frac{\partial F_\varphi}{\partial\varphi}, \quad (215)$$

$$\nabla^2 V = \frac{1}{r^2}\frac{\partial}{\partial r}\left(r^2\frac{\partial V}{\partial r}\right) + \frac{1}{r^2\sin\theta}\frac{\partial}{\partial\theta}\left(\sin\theta\frac{\partial V}{\partial\theta}\right) +$$
$$+ \frac{1}{r^2\sin^2\theta}\frac{\partial^2 V}{\partial\varphi^2}. \quad (216)$$

The Curl Operator

We obtain the expression for (curl $\mathbf{F}$)$_\lambda$ by applying eqn. (186) to the rectangle $PSWT$, and can then readily find (curl $\mathbf{F}$)$_\mu$ and (curl $\mathbf{F}$)$_\nu$. It is clear that since $PSWT$ is rectangular

$$\delta S = PS \times PT = h_\mu h_\nu\, \delta\mu\, \delta\nu \quad (217)$$

from eqn. (202). To calculate $\int \mathbf{F}.d\mathbf{r}$ for $PSWT$, we use a similar approach to that employed in proving the first part of Stokes' theorem in §7.3, and begin by evaluating $\int \mathbf{F}.d\mathbf{r}$ for the sides PS and WT; we assume that $PSWT$ is traversed in an anticlockwise direction when viewed from inside the parallelepiped. It then follows in complete analogy with the derivation of eqn. (176) that

$$\int_{\substack{PS \\ \text{and } WT}} \mathbf{F}.d\mathbf{r} = -[\partial(F_\mu PS)/\partial v]\,\delta v$$

$$= -[\partial(h_\mu F_\mu)/\partial v]\,\delta\mu\,\delta v, \quad (218)$$

since $PS = h_\mu \, \delta\mu$ and $\delta\mu$ is independent of v†. Similarly, if we calculate $\int \mathbf{F} . \, d\mathbf{r}$ for the pair of sides PT and SW we obtain

$$\int_{\substack{TP \\ \text{and } SW}} \mathbf{F} . \, d\mathbf{r} = + [\partial(h_v F_v)/\partial\mu] \, \delta\mu \, \delta v, \tag{219}$$

the difference in sign between eqns. (218) and (219) being for the same reason as that given below eqn. (179). Adding the expressions (218) and (219) gives

$$\oint_{PSWT} \mathbf{F} . \, d\mathbf{r} = \{[\partial(h_v F_v)/\partial\mu] - [\partial(h_\mu F_\mu)/\partial v]\} \, \delta\mu \, \delta v,$$

and dividing this by expression (217) for δS before taking the limit as $\delta\lambda, \, \delta\mu, \, \delta v \to 0$, yields finally

$$(\text{curl } \mathbf{F})_\lambda = \lim_{\delta s \to 0} \frac{\oint \mathbf{F} . \, d\mathbf{r}}{\delta S} = \frac{1}{h_\mu h_v} \left(\frac{\partial(h_v F_v)}{\partial\mu} - \frac{\partial(h_\mu F_\mu)}{\partial v} \right). \tag{220}$$

Similarly, by using this approach for the rectangles $PQUT$ and $PQRS$, we obtain

$$(\text{curl } \mathbf{F})_\mu = \frac{1}{h_\lambda h_v} \left[\frac{\partial(h_\lambda F_\lambda)}{\partial v} - \frac{\partial(h_v F_v)}{\partial\lambda} \right],$$

$$(\text{curl } \mathbf{F})_v = \frac{1}{h_\lambda h_\mu} \left[\frac{\partial(h_\mu F_\mu)}{\partial\lambda} - \frac{\partial(h_\lambda F_\lambda)}{\partial\mu} \right]. \tag{221}$$

Thus in terms of the unit vectors $\boldsymbol{\alpha}$, $\boldsymbol{\beta}$, $\boldsymbol{\gamma}$ defined earlier

$$\text{curl } \mathbf{F} = \frac{\boldsymbol{\alpha}}{h_\mu h_v} \left[\frac{\partial(h_v F_v)}{\partial\mu} - \frac{\partial(h_\mu F_\mu)}{\partial v} \right] +$$

$$+ \frac{\boldsymbol{\beta}}{h_\lambda h_v} \left[\frac{\partial(h_\lambda F_\lambda)}{\partial v} - \frac{\partial(h_v F_v)}{\partial\lambda} \right] + \frac{\boldsymbol{\gamma}}{h_\lambda h_\mu} \left[\frac{\partial(h_\mu F_\mu)}{\partial\lambda} - \frac{\partial(h_\lambda F_\lambda)}{\partial\mu} \right]. \tag{222}$$

---

† It should be noticed that in this calculation, we have assumed that $PSWT$ is a perfect rectangle for the purpose of calculating its area, while the result (217) contains a term $\partial h_\mu/\partial v$ corresponding to PS not being exactly equal to WT. The reasons for this are similar to those given in the footnote to eqn. (208): the term arising from $\partial h_\lambda/\partial v$ on the R.H.S. of eqn. (217) is of the same order of magnitude as the term arising from $\partial F_\mu/\partial v$, while on the other hand the deviation from equality of PS and WT gives an additional term to the area, whose ratio to the true area tends to zero as $\delta\lambda, \, \delta\mu, \, \delta v \to 0$.

The result (222) may be immediately applied to both cylindrical and spherical polar coordinates, when we obtain from eqns. (203) and (204) that for cylindrical polars

$$\text{curl } \mathbf{F} = \mathscr{I}\left[\frac{1}{\rho}\frac{\partial F_z}{\partial \varphi} - \frac{\partial F_\varphi}{\partial z}\right] + \mathscr{J}\left[\frac{\partial F_\rho}{\partial z} - \frac{\partial F_z}{\partial \rho}\right] +$$
$$+ \mathscr{K}\left[\frac{1}{\rho}\frac{\partial(\rho F_\varphi)}{\partial \rho} - \frac{1}{\rho}\frac{\partial F_\rho}{\partial \varphi}\right], \quad (223)$$

and for spherical polars

$$\text{curl } \mathbf{F} = \mathbf{I}\frac{1}{r\sin\theta}\left[\frac{\partial(\sin\theta\, F_\varphi)}{\partial\theta} - \frac{\partial F_\theta}{\partial\varphi}\right] +$$
$$+ \mathbf{J}\frac{1}{r}\left[\frac{1}{\sin\theta}\frac{\partial F_r}{\partial\varphi} - \frac{\partial(rF_\varphi)}{\partial r}\right] + \mathbf{K}\frac{1}{r}\left[\frac{\partial(rF_\theta)}{\partial r} - \frac{\partial F_r}{\partial\theta}\right]. \quad (224)$$

Worked example

Use spherical polar coordinates to prove div $(\mathbf{r}/r^3) = 0$.

Ans. If $\mathbf{F} = \mathbf{r}/r^3$ then in spherical polar coordinates $F_r = r^{-2}$, $F_\theta = 0 = F_\varphi$. Hence from eqn. (215)

$$\text{div } \mathbf{F} = \frac{1}{r^2}\frac{\partial}{\partial r}\left(r^2 \times \frac{1}{r^2}\right)$$

since $F_\theta = F_\varphi = 0$. Thus div $\mathbf{F} = 0$.

Exercise

Use spherical polar coordinates to prove that
(a) $\nabla^2(1/r) = 0$,
(b) curl $f(r)\mathbf{r} = 0$,
(c) div $f(r)\mathbf{r} = 3f + r\,\mathrm{d}f/\mathrm{d}r$.

9

An Application of Vector Analysis—Electrical Theory

9.1 ELECTROSTATIC FIELD AND POTENTIAL

The fundamental experimental law of electrostatics states that two positive point charges in free space repel each other with a force directed along the line joining the charges; the force is proportional to the product of the charges and inversely proportional to the square of their separation. Thus if the charges are q_1 and q_2 situated at O and P respectively, as shown in Fig. 65, then the force $\mathbf{F}$ exerted by q_1 on q_2 is given by

$$\mathbf{F} = q_1 q_2 \mathbf{r}/\varepsilon_0 r^3, \tag{225}$$

where $\mathbf{r} = \overrightarrow{OP}$ and ε_0^{-1} is a constant of proportionality whose value depends on the system of units employed. We define the *electrical field* $\mathbf{E}$ at a point P due to a system of charges as the force acting on unit positive charge placed at P, and it follows from eqn. (225) that the field $\mathbf{E}$ at P due to a point charge q at O is given by

$$\mathbf{E}(\mathbf{r}) = q\mathbf{r}/\varepsilon_0 r^3. \tag{226}$$

It is clear from eqn. (226) that the electrical field $\mathbf{E}(\mathbf{r})$ constitutes a vector field, since for a fixed charge q, eqn. (226) allots a vector $\mathbf{E}$ to every point P in space.

Electrostatic Potential

We now proceed to show that the electrical field $\mathbf{E(r)}$ may be derived from a scalar field $V\mathbf{(r)}$ via the relation

$$\mathbf{E} = -\operatorname{grad} V.\dagger \qquad (227)$$

It has been shown in §7.3 that a necessary and sufficient condition for this to be so is that curl $\mathbf{E} = 0$. Now, from eqn. (226) it follows that

$$\operatorname{curl} \mathbf{E} = (q/\varepsilon_0) \operatorname{curl} (\mathbf{r}/r^3), \qquad (227a)$$

FIG. 65.

and this is zero since curl $f(r)\mathbf{r} = 0$ for all $f(r)$ as considered in exercise (*b*) at the end of §8.2.

Since $\mathbf{E} = -\operatorname{grad} V$, it follows that $\mathbf{E}$ is a conservative field, and so $\int_A^B \mathbf{E}.d\mathbf{r}$ is independent of the path taken between A and B. This result has a simple physical interpretation since we have shown in §6.1 that for a force field $\mathbf{F}$, $\int_A^B \mathbf{F}.d\mathbf{r}$ is the work done *by* the field $\mathbf{F}$ in moving from A to B. Hence $\int_A^B \mathbf{E}.d\mathbf{r}$ is the work done by the field in taking a unit positive charge from A to B, and the above result shows that this is independent of the path taken between A and B. In fact

$$\int_A^B \mathbf{E}.d\mathbf{r} = -\int_A^B \operatorname{grad} V.d\mathbf{r} = \int_B^A dV = V_A - V_B \qquad (228)$$

from eqns. (87) and (227). Now, since $\int_A^B \mathbf{E}.d\mathbf{r}$ is the work done by the field in taking unit positive charge from A to B, it must equal the potential energy of the charge at A minus that at B. Hence by comparison with the R.H.S. of eqn. (228), it follows that V may be identified with the potential energy of a unit

† The minus sign on the R.H.S. of eqn. (227) is in order that V may be later given a specific physical interpretation.

positive charge in the electrostatic field; V is termed the *electrostatic potential* of the field. This potential energy is always arbitrary to the extent of a single additive constant, and if this constant is fixed by supposing V at infinity to be zero, then $V(\mathbf{r})$ will represent physically the work done *on* unit positive charge *against* the field in taking it from infinity to the point with position vector $\mathbf{r}$. It is readily seen that for the point charge with which we are concerned, $V(\mathbf{r})$ is given by

$$V(\mathbf{r}) = +q/\varepsilon_0 r, \tag{229}$$

since working in spherical polar coordinates, we have from eqn. (206a)

$$-\operatorname{grad} V = -\mathbf{I}(q/\varepsilon_0)\, \mathrm{d}(r^{-1})/\mathrm{d}r$$
$$= \mathbf{I}(q/\varepsilon_0)r^{-2} = q\mathbf{r}/\varepsilon_0 r^3 = \mathbf{E}(\mathbf{r})$$

as V is independent of θ and φ; also $V(\infty) = 0$.

Assembly of Point Charges

So far we have considered the field and potential due to a *single* point charge. Let us suppose now that there exists a system of N such charges $q_1, q_2, \ldots, q_p, \ldots, q_N$ situated respectively at points $P_1, P_2, \ldots, P_N$ with position vectors $\mathbf{r}_1, \mathbf{r}_2, \ldots, \mathbf{r}_N$. Then if $P(\mathbf{r})$ is a general point in space, it follows from eqn. (225) that the field $\mathbf{E}_1$ at P due to q_1 is given by

$$\mathbf{E}_1 = q_1(\mathbf{r} - \mathbf{r}_1)/\varepsilon_0 |\mathbf{r} - \mathbf{r}_1|^3$$

and so the total field $\mathbf{E}$ due to all the charges is given by

$$\mathbf{E} = \sum_{n=1}^{N} q_n(\mathbf{r} - \mathbf{r}_n)/\varepsilon_0 |\mathbf{r} - \mathbf{r}_n|^3. \tag{230}$$

To show that here also a scalar field V exists such that $\mathbf{E} = -\operatorname{grad} V$, we must show that $\operatorname{curl} \mathbf{E} = 0$. To do this, we let $\mathbf{r} - \mathbf{r}_n = \mathbf{S}_n$, and then

$$\mathbf{E} = \sum_{n=1}^{N} \mathbf{E}_n, \quad \text{where } \mathbf{E}_n = q_n \mathbf{S}_n/\varepsilon_0 S_n^3.$$

Now, applying the curl operator here means taking the curl with respect to $\mathbf{r}$ as the independent variable. However, since the $\mathbf{S}_n$ only differ from $\mathbf{r}$ by a fixed vector $\mathbf{r}_n$ (for given n), the curl may also be taken with respect to any of the $\mathbf{S}_n$. Hence

$$\operatorname{curl}_r \mathbf{E}_n = \operatorname{curl}_{S_n} \mathbf{E}_n = (q_n/\varepsilon_0) \operatorname{curl}_{S_n} (\mathbf{S}_n/S_n^3) = 0$$

by the same argument as was used earlier in connection with eqn. $(227a)$, and so

$$\operatorname{curl} \mathbf{E} = \sum_{n=1}^{N} \operatorname{curl} \mathbf{E}_n = 0$$

showing that $V(\mathbf{r})$ exists. The discussion of $\int_A^B \mathbf{E} \cdot d\mathbf{r}$ proceeds exactly as before in the case of the single point charge, and again $V(\mathbf{r})$, the electrostatic potential, is the potential energy of a unit charge at $P(\mathbf{r})$. In this case, an explicit form for $V(\mathbf{r})$ is given by

$$V(\mathbf{r}) = \sum_{n=1}^{N} q_n/\varepsilon_0 |\mathbf{r} - \mathbf{r}_n|. \tag{231}$$

It is left as an exercise for the student to show that this is correct.

Continuous Charge Distributions

So far we have considered discrete point charges. However, in many contexts one is concerned with a continuous charge distribution which is defined by a charge density $\rho(\mathbf{r})$ (itself a scalar field) such that the total charge in a small volume $\delta\tau$ is given by $\rho(\mathbf{r}) \, \delta\tau$; under these conditions, the total charge in a volume τ is clearly $\int_\tau \rho(\mathbf{r}) \, d\tau$. If we suppose the volume τ to be completely divided into N elemental volumes $\delta\tau_n(1 \leqq n \leqq N)$, the charge q_n in $\delta\tau_n$ is given by $q_n = \rho(\mathbf{r}) \, \delta\tau_n$, and so applying the results (230) and (231) to the whole volume, we find the total field $\mathbf{E}$ and total potential V to be given by

$$\mathbf{E}(\mathbf{R}) = \int_\tau \left[\rho(\mathbf{r})(\mathbf{R} - \mathbf{r})/\varepsilon_0 \left|\mathbf{R} - \mathbf{r}\right|^3\right] d\tau, \qquad (232)$$

$$V(\mathbf{R}) = \int_\tau \left[\rho(\mathbf{r})/\varepsilon_0 \left|\mathbf{R} - \mathbf{r}\right|\right] d\tau. \qquad (233)$$

Care should be taken in these results to distinguish between the current integration variable $\mathbf{r}$ and the vector position $\mathbf{R}$ of the point at which $\mathbf{E}$ and V are being observed.

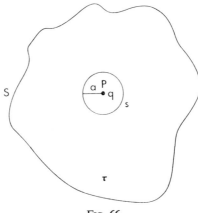

Fig. 66.

Finally, it should be noticed that since the charge in an element $\delta\tau$ is $\rho\, \delta\tau$, the force $\delta\mathbf{F}$ on such an element is given by

$$\delta\mathbf{F} = \mathbf{E}\rho\, \delta\tau. \qquad (233a)$$

9.2 GAUSS' THEOREM

Consider a point charge q at point P inside a closed surface S as shown in Fig. 66. Then if $\mathbf{E}(\mathbf{r})$ is the electric field due to q, Gauss' theorem states that

$$\int \mathbf{E}.d\mathbf{S} = 4\pi q/\varepsilon_0. \qquad (234)$$

We prove the theorem by taking a small sphere s of radius a and centre P, when we have

$$\int_{S+s} \mathbf{E}.d\mathbf{S} = \int_S \mathbf{E}.d\mathbf{S} + \int_s \mathbf{E}.d\mathbf{S} = \int_\tau \text{div } \mathbf{E} \, d\tau \quad (235)$$

where τ is the volume between S and s. Now,

$$\text{div } \mathbf{E} = (q/\varepsilon_0) \text{ div } (\mathbf{r}/r^3) = 0 \text{ for } r \neq 0$$

as shown at the end of §8.2, and hence

$$\int_S \mathbf{E}.d\mathbf{S} = - \int_s \mathbf{E}.d\mathbf{S}. \quad (236)$$

On s,

$$\mathbf{E}.d\mathbf{S} = \mathbf{E}.\mathbf{n} \, dS = dS(q/\varepsilon_0)(\mathbf{r}.\mathbf{n})/r^3 \quad \text{from eqn. (226)}$$

$$= dS(q/\varepsilon_0)[\mathbf{r}.(\mathbf{r}/a)]/a^3 = -dS(q/\varepsilon_0)/a^2$$

since $|\mathbf{r}| = a$, and $\mathbf{n} = -\mathbf{r}/a$ as $\mathbf{n}$ is taken *out* of τ and therefore towards P. Substituting this into eqn. (236) gives

$$\int_S \mathbf{E}.d\mathbf{S} = (q/\varepsilon_0)a^{-2} \int_s \, dS = 4\pi q/\varepsilon_0 \quad (237)$$

since $\int_s \, dS = 4\pi a^2$. It also follows from this approach that if q lies outside the closed surface S, then $\int_S \mathbf{E}.d\mathbf{S} = 0$, since we have

$$\int_S \mathbf{E}.d\mathbf{S} = \int_\tau \text{div } \mathbf{E} \, d\tau = 0$$

as div $\mathbf{E} = 0$ for $r \neq 0$. This result is true as long as $r \neq 0$; that is, as long as q is outside S.

Suppose now that there are a set of changes $q_1, q_2, \ldots, q_n$ situated at points $P_1, P_2, \ldots, P_n$ all lying inside S. Then if $\mathbf{E}_n$ is the field due to charge q_n it follows from eqn. (234) that

$$\int \mathbf{E}_n.d\mathbf{S} = (4\pi/\varepsilon_0)q_n$$

and summing both sides of this equation for all n gives

$$\int \mathbf{E}.d\mathbf{S} = \sum_{n=1}^{N} \int \mathbf{E}_n.d\mathbf{S} = (4\pi/\varepsilon_0) \sum_n q_n = 4\pi q/\varepsilon_0, \qquad (238)$$

where $\mathbf{E}(= \sum_n \mathbf{E}_n)$ and $q(= \sum_n q_n)$ is the total electric field and charge respectively. Equation (238) shows that Gauss' theorem remains true as long as q is the total charge inside S; this charge need not be concentrated at a single point. It also follows from the remarks after eqn. (237) that for any charge distribution lying outside S, $\int \mathbf{E}.d\mathbf{S} = 0$.

9.3 POISSON'S AND LAPLACE'S EQUATIONS

Suppose now that we have a continuous charge distribution with charge density $\rho(\mathbf{r})$. Then applying Gauss' theorem to an arbitrary volume τ enclosed by a surface S and containing charge $q = \int_\tau \rho d\tau$ we have

$$(4\pi/\varepsilon_0) \int \rho \, d\tau = \int \mathbf{E}.d\mathbf{S} = \int \text{div } \mathbf{E} \, d\tau$$

from eqn. (160). It follows that

$$\int [\text{div } \mathbf{E} - (4\pi\rho/\varepsilon_0)] \, d\tau = 0$$

for an *arbitrary* volume τ, and hence

$$\text{div } \mathbf{E} = 4\pi\rho/\varepsilon_0. \qquad (239)$$

Substituting from eqn. (227), we obtain *Poisson's* equation

$$\nabla^2 V = \text{div grad } V = -4\pi\rho/\varepsilon_0. \qquad (240)$$

If $\rho = 0$, corresponding to the absence of charge, we obtain *Laplace's* equation

$$\nabla^2 V = 0. \qquad (241)$$

One of the most important problems of electrostatics is to solve Poisson's equation for V, when ρ is given together with suitable boundary conditions for V. It is clear that this equation is a second-order equation in three independent variables; its solution can become very involved and we shall therefore not consider it in detail here. However, we shall now prove the following important uniqueness theorem: *V is determined uniquely inside a closed surface S, if it satisfies Poisson's equation inside S, and in addition V is given at all points on S.* To prove this, suppose that there are two potentials U and V satisfying the given conditions; i.e. $\nabla^2 U = \nabla^2 V = 4\pi\rho$ inside S, $U = V$ on S. Then if $W = U - V$, we see from the above conditions on U and V that

$$\nabla^2 W = 0 \quad \text{inside } S \text{ and} \quad W = 0 \text{ on } S. \qquad (242)$$

Let us now apply Green's first theorem (eqn. (164)) to W, when we obtain

$$\int W \operatorname{grad} W.\mathrm{d}\mathbf{S} = \int |\operatorname{grad} W|^2 \, \mathrm{d}\tau + \int W \nabla^2 W \, \mathrm{d}\tau. \qquad (243)$$

The L.H.S. of this equation is zero since $W = 0$ on S, and so also is the second term on the R.H.S. since $\nabla^2 W = 0$ inside S. Thus we obtain

$$\int |\operatorname{grad} W|^2 \, \mathrm{d}\tau = 0$$

and since the integrand, being a square, must be $\geqq 0$ it follows that $\operatorname{grad} W = 0$. Since $\operatorname{grad} W = 0$ it follows that W is constant inside S, and since it is zero *on* S, it must be zero everywhere. As $W = 0$ everywhere, it follows that $U = V$. Hence since any two solutions of the problem are identical, it follows that the given conditions determine a unique solution.

9.4 ENERGY OF THE ELECTROSTATIC FIELD

Suppose that the whole of space is filled with electric charge of density $\rho(\mathbf{r})$, giving rise to an electric field $\mathbf{E}(\mathbf{r})$ and corresponding potential $V(\mathbf{r})$. What is the energy of this field?

To obtain this we shall build up the field by bringing charge from infinity, so that at any stage of the process the charge density at **r** is $\lambda\rho(\mathbf{r})$, where λ ranges from 0 at the beginning of the process to 1 at the end of it. When the charge density at **r** is $\lambda\rho(\mathbf{r})$, the potential there will be $\lambda V(\mathbf{r})$, and so if λ is increased by $d\lambda$, the increase in energy of an elemental volume $\delta\tau$ at **r** will be the potential at **r** $\times$ the increase in charge

$$= \lambda V(\mathbf{r}) \times d\lambda\rho(\mathbf{r}) \, d\tau.$$

Hence the total increase in energy dW for an increase $d\lambda$ in λ is given by

$$dW = \lambda \, d\lambda \int_\tau V\rho \, d\tau.$$

Thus the total energy W at the end of the building-up process is given by

$$W = \int_\tau V\rho \, d\tau \int_0^1 \lambda \, d\lambda = \tfrac{1}{2} \int_\tau V\rho \, d\tau, \tag{244}$$

and in any real situation it may be assumed that V and $\rho \to 0$ sufficiently fast as $r \to \infty$ for the integral on the R.H.S. of eqn. (244) to converge. We may obtain W in terms of **E** by making use of eqn. (239) to yield

$$W = (\varepsilon_0/8\pi) \int V \operatorname{div} \mathbf{E} \, d\tau \tag{245}$$

$$= (\varepsilon_0/8\pi)[\int \operatorname{div}(\mathbf{E}V) \, d\tau - \int \mathbf{E}.\operatorname{grad} V \, d\tau] \tag{246}$$

employing eqn. (102). Now the first term on the R.H.S. of eqn. (246) involves

$$\int \operatorname{div}(\mathbf{E}V) \, d\tau = \int_s V\mathbf{E}.d\mathbf{S} \tag{247}$$

where the surface integral is to be taken over the boundary surface at infinity, and it may be shown that if V and ρ tend to zero at such a rate that the R.H.S. of eqn. (244) converges as $r \to \infty$, then $\int V\mathbf{E}.d\mathbf{S} \to 0$ as $r \to \infty$. Thus the first term on the R.H.S. of eqn. (246) is zero, and substituting $\mathbf{E} = -\text{grad } V$ in the second term gives

$$W = (\varepsilon_0/8\pi) \int E^2 \, d\tau. \tag{248}$$

It is therefore possible to define an energy density $w(\mathbf{r}) = \varepsilon_0 E^2/8\pi$ in the sense that $W = \int w \, d\tau$.

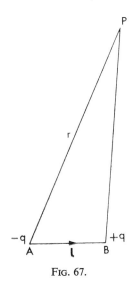

FIG. 67.

9.5 DIPOLES

Consider two charges $-q$ and $+q$ at A and B as shown in Fig. 67, such that $AB \ll AP$, where P is the point at which observations of fields and potential are being made. Then if $\overrightarrow{AB} = \mathbf{l}$ and $AP = r$, the potential V at P is given by

$$V = +q[BP^{-1} - AP^{-1}]. \tag{249}$$

Now, since $AP^{-1} = r^{-1}$,

$$BP^{-1} - AP^{-1} \approx \delta(r^{-1}) \approx \delta\mathbf{r} . \text{grad } (r^{-1}) \text{ from eqn. (87).}$$

Since $\delta\mathbf{r} = \mathbf{l}$, this gives

$$V \approx q\mathbf{l}.\text{grad}\,(r^{-1}) \tag{250}$$

and this relation becomes rigorously correct in the limit as $l \to 0$ and $q \to \infty$ with $q\mathbf{l}$ retaining its same (finite) value. If, then, we define the dipole moment $\mathbf{p}$ of the two charges by

$$\mathbf{p} = q\mathbf{l} \tag{251}$$

we have from eqn. (250)

$$V = \mathbf{p}.\text{grad}_A(r^{-1}). \tag{252}$$

We have here written subscript A to the grad, since the method by which it was obtained shows that the grad is to be taken at A and not at P. If the grad is taken at P, we have

$$\text{grad}_P\,(r^{-1}) = -\text{grad}_A\,(r^{-1})$$

and so

$$V = -\mathbf{p}.\text{grad}_P\,(r^{-1}) = \mathbf{p}.\mathbf{r}/r^3$$

from eqn. (91a).
The field $\mathbf{E}$ at P is given by

$$\mathbf{E} = -\text{grad}\,V = -\text{grad}\,(\mathbf{p}.\mathbf{r}/r^3) = \frac{3(\mathbf{p}.\mathbf{r})\mathbf{r}}{r^5} - \frac{\mathbf{p}}{r^3}. \tag{253}$$

It is left to the reader to verify the final step in eqn. (253).

Potential Energy and Force on the Dipole in an External Field

If the dipole is placed in an external field characterised by potential $V(\mathbf{r})$, then the potential energy of $-q$ and $+q$ are $-qV_A$ and $+qV_B$ respectively. Hence the potential energy U of the dipole is given by

$$U = +q(V_B - V_A) = q\mathbf{l}.\text{grad}\,V$$

from eqn. (87) with $d\mathbf{r} = \mathbf{l}$. Thus, since $q\mathbf{l} = \mathbf{p}$ and $\mathbf{E} = -\text{grad } V$, we have finally

$$U = -\mathbf{p}.\mathbf{E}. \tag{254}$$

Now the force $\mathbf{F}$ on the dipole is given by

$$\mathbf{F} = -\text{grad } U = \text{grad } (\mathbf{p}.\mathbf{E})$$

from eqn. (254). Employing eqn. (101), we then obtain

$$\mathbf{F} = (\mathbf{p}.\text{grad}) \mathbf{E} \tag{255}$$

since $\mathbf{p}$ is constant, and curl $\mathbf{E} = 0$.

In order to obtain the moment about an external point P of the forces acting on the dipole, we note first that these forces may be considered to consist of a couple arising from equal and opposite forces $\pm q\mathbf{E}$ acting at A and B, together with a resultant force $\mathbf{F}$ (eqn. (255)) acting on the dipole. The magnitude $\mathbf{m}$ of the couple is given by

$$\mathbf{m} = \mathbf{l} \wedge q\mathbf{E} = \mathbf{p} \wedge \mathbf{E}$$

(see eqn. (33)), and the moment $\boldsymbol{\mu}$ of $\mathbf{F}$ about P is given by

$$\boldsymbol{\mu} = \mathbf{r} \wedge \mathbf{F} = \mathbf{r} \wedge (\mathbf{p}.\text{grad}) \mathbf{E}.$$

Hence the total moment $\mathbf{M}$ about P of the forces acting on the dipole is

$$\mathbf{M} = \mathbf{m} + \boldsymbol{\mu} = \mathbf{p} \wedge \mathbf{E} + \mathbf{r} \wedge (\mathbf{p}.\text{grad}) \mathbf{E}.$$

9.6 CONDUCTORS AND INSULATORS

So far we have discussed the theory of electrostatics in empty space. Now let us consider the electric fields existing inside a material body. In general we may divide substances into two classes: conductors and insulators. In conductors the electrons which form part of the internal atomic structure are free to move under the action of external fields, and hence if such a field is applied, electrons will move from one part of the body to another in order to set up an internal charge density $\rho(\mathbf{r})$, which itself gives rise to a field equal and opposite

M

to the external field. When this has been achieved no net field exists inside the conductor, and thus the electron movement stops. The equilibrium situation for a conductor is therefore that the total internal field is zero corresponding to a constant potential throughout the conductor.

For insulators, on the other hand, the internal atomic structure allows no net flow of electrons through the medium. However, an external field *does* produce a slight separation between positive and negative charges in the atom, and this in turn gives rise to an internal dipole moment which is usually specified by a *polarisation* **P**, defined as being the dipole moment for unit volume. It is found by experiment that **P** is proportional to the external field **E**, and thus

$$\mathbf{P} = k\mathbf{E} \tag{256}$$

where the proportionality constant k is known as the dielectric susceptibility.

Potential Due to Polarisation

We can now show that the electrical effect of a polarisation **P** inside an insulator is equivalent to the effect of a charge density ρ given by $\rho = -\operatorname{div} \mathbf{P}$, together with a certain surface charge density. To show this, consider the potential V due to the polarisation of a given volume τ of insulator. It follows immediately from eqn. (252) and the definition of **P** that

$$V = \int_\tau \mathbf{P} . \operatorname{grad} (r^{-1}) \, d\tau. \tag{257}$$

Now, from eqn. (102) we have

$$\operatorname{div} (\mathbf{P}r^{-1}) - r^{-1} \operatorname{div} \mathbf{P} = \mathbf{P} . \operatorname{grad} (r^{-1})$$

and thus eqn. (257) yields

$$V = \int_\tau \operatorname{div} (\mathbf{P}r^{-1}) \, d\tau - \int r^{-1} \operatorname{div} \mathbf{P} \, d\tau$$

$$= \int r^{-1} \mathbf{P} . d\mathbf{S} - \int_\tau r^{-1} \operatorname{div} \mathbf{P} \, d\tau. \tag{258}$$

Since the potential V' due to a charge density ρ is given by $V' = \int r^{-1} \rho \, d\tau$, we see that the second term on the R.H.S. of eqn. (258) corresponds to the potential due to a charge density equal to $-\text{div } \mathbf{P}$, while the first term is the potential due to a surface charge density equal to P_n, the normal component of $\mathbf{P}$.

Modification of Gauss' Theorem due to Polarisation

We obtained earlier the result $\text{div } \mathbf{E} = 4\pi\rho'/\varepsilon_0$, where ρ' is the net charge density. In the presence of a polarised insulator, this result will remain correct if ρ' is now the sum of the actual charge density ρ and the " polarisation charge density " $-\text{div } \mathbf{P}$. Thus we obtain $\text{div } \mathbf{E} = (4\pi/\varepsilon_0)(\rho - \text{div } \mathbf{P})$, and therefore

$$\text{div } [\mathbf{E} + (4\pi\mathbf{P}/\varepsilon_0)] = 4\pi\rho/\varepsilon_0. \tag{259}$$

Defining now a vector $\mathbf{D} = \mathbf{E} + (4\pi\mathbf{P}/\varepsilon_0)$, we obtain

$$\text{div } \mathbf{D} = 4\pi\rho/\varepsilon_0 \tag{260}$$

which is the required modification of eqn. (239) brought about by the presence of the dielectric. The above definition of $\mathbf{D}$ may be simplified by use of eqn. (256) which yields

$$\mathbf{D} = \mathbf{E} + (4\pi k/\varepsilon_0)\mathbf{E} = [1 + (4\pi k/\varepsilon_0)]\mathbf{E} = \varepsilon\mathbf{E} \tag{261}$$

where the *dielectric constant* ε is defined by $\varepsilon = 1 + (4\pi k/\varepsilon_0)$. Making use of this result, and assuming ε to be independent of position, eqn. (260) may be written

$$\text{div } \mathbf{E} = 4\pi\rho/\varepsilon\varepsilon_0,$$

while Poisson's equation in the presence of an insulator becomes

$$\nabla^2 V = -4\pi\rho/\varepsilon\varepsilon_0.$$

9.7 ELECTRIC CURRENT

So far we have been concerned with the effects due to electric charges and charge distributions which do not alter with time. However, if we apply an electric field along a

M*

conductor, and prevent a static charge distribution from building up by completing an electrical circuit, then it is possible to have a continual flow of charge around the circuit. This flow is characterised by a *current density* $\mathbf{J}$ defined by

$$\mathbf{J} = \rho\mathbf{v} \tag{262}$$

where ρ is the charge density and $\mathbf{v}$ is the velocity of motion of the electric charge.

Let us consider a closed surface S through which charge is flowing. Then the total charge flowing out of the surface per unit time is

$$\int_s \rho v_n \, \mathrm{d}S = \int \rho\mathbf{v}.\mathrm{d}\mathbf{S} = \int \mathbf{J}.\mathrm{d}\mathbf{S}$$

where v_n is the resolute of velocity perpendicular to the surface. This must equal the rate of decrease of charge in the volume τ enclosed by S, which is given by

$$-\partial(\int_\tau \rho \, \mathrm{d}\tau)/\partial t = -\int (\partial\rho/\partial t) \, \mathrm{d}\tau.$$

Equating these two expressions yields

$$\int_\tau (\partial\rho/\partial t) \, \mathrm{d}\tau = -\int \mathbf{J}.\mathrm{d}\mathbf{S} = -\int \mathrm{div} \, \mathbf{J} \, \mathrm{d}\tau$$

by the divergence theorem. Hence we have

$$\int_\tau [(\partial\rho/\partial t) + \mathrm{div} \, \mathbf{J}] \, \mathrm{d}\tau = 0$$

and since this is true for *any* volume τ, it follows that

$$\mathrm{div} \, \mathbf{J} + (\partial\rho/\partial t) = 0. \tag{263}$$

For steady currents $(\partial\rho/\partial t) = 0$ and then $\mathrm{div} \, \mathbf{J} = 0$.

In practice we are often concerned with the steady passage of charge along wires, which are effectively conductors of very small cross-section (compared with their length). In this case, our interest is mainly in the total current I, defined by

$$I = \int_s \mathbf{J}.\mathrm{d}\mathbf{S} \tag{264}$$

where S is the cross-section of the wire.

9.8   MAGNETIC EFFECTS OF A CURRENT

Consider two wire circuits C_1 and C_2 as shown in Fig. 68 carrying currents I_1 and I_2 respectively. Then it is found by experiment that a force exists between the two circuits, which may be described in the following way. Let $\delta\mathbf{r}_1$, $\delta\mathbf{r}_2$ be vector elements of the two circuits, separated by a displacement $\mathbf{r}$. Then the force $\delta\mathbf{F}$ on $\delta\mathbf{r}_1$ due to the circuit C_2 is given by

$$\delta\mathbf{F} = \mu_0 I_1 I_2 \, \delta\mathbf{r}_1 \wedge \oint_{C_2} (\mathrm{d}\mathbf{r}_2 \wedge \mathbf{r})/r^3,$$

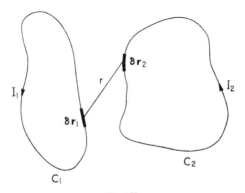

FIG. 68.

where μ_0 is a constant of proportionality. It is convenient to introduce now the magnetic field $\mathbf{H}$ at $\delta\mathbf{r}_1$ due to the circuit C_2; this is defined by

$$\mathbf{H} = I_2 \int_{C_2} \mathrm{d}\mathbf{r}_2 \wedge \mathbf{r}/r^3. \tag{265}$$

In terms of $\mathbf{H}$ it then follows that the force on $\delta\mathbf{r}_1$ due to the circuit C_2 is given by

$$\delta\mathbf{F} = \mu_0 I_1 \, \delta\mathbf{r}_1 \wedge \mathbf{H}. \tag{266}$$

9.9   MAGNETIC VECTOR POTENTIAL

We first show that the magnetic field $\mathbf{H}$ (eqn. (265)) may be derived from a vector potential $\mathbf{A}$ via the relation $\mathbf{H} = \operatorname{curl} \mathbf{A}$. To prove this, we consider

$$
\begin{aligned}
\operatorname{div} \mathbf{H} &= I_2 \oint_{C_2} \operatorname{div}_r [d\mathbf{r}_2 \wedge \mathbf{r}/r^3] \\
&= I_2 \oint_{C_2} [(\mathbf{r}/r^3).\operatorname{curl}_r d\mathbf{r}_2 - d\mathbf{r}_2.\operatorname{curl}_r (\mathbf{r}/r^3)]
\end{aligned}
\tag{267}
$$

by eqn. (103). Now $\operatorname{curl}_r d\mathbf{r}_2 = 0$, since $d\mathbf{r}_2$ is independent of $\mathbf{r}$, and $\operatorname{curl}_r (\mathbf{r}/r^3) = 0$ from the discussion after eqn. (227a). Hence the integrand in eqn. (267) is zero, and so

$$
\operatorname{div} \mathbf{H} = 0.
\tag{267a}
$$

It follows then from the discussion in §5.6 that a vector $\mathbf{A}$ exists such that $\mathbf{H} = \operatorname{curl} \mathbf{A}$. It was mentioned in §5.6 that the relation $\mathbf{H} = \operatorname{curl} \mathbf{A}$ alone does not determine $\mathbf{A}$ uniquely, but that if we impose the condition $\operatorname{div} \mathbf{A} = 0$, a unique vector potential $\mathbf{A}$ *does* result. For the present case, the vector potential $\mathbf{A}$ satisfying $\operatorname{div} \mathbf{A} = 0$ is given by

$$
\mathbf{A} = I_2 \oint_{C_2} d\mathbf{r}_2/r
\tag{268}
$$

and this result we now proceed to prove. Consider

$$
\begin{aligned}
\operatorname{curl} \mathbf{A} &= I_2 \oint_{C_2} \operatorname{curl}_r (d\mathbf{r}_2/r) \\
&= I_2 \oint_{C_2} [r^{-1} \operatorname{curl}_r d\mathbf{r}_2 + \operatorname{grad}_r(r^{-1}) \wedge d\mathbf{r}_2]
\end{aligned}
$$

from eqn. (104). Now, $\operatorname{curl}_r d\mathbf{r}_2 = 0$ since $d\mathbf{r}_2$ is independent of $\mathbf{r}$, and $\operatorname{grad}_r (r^{-1}) = -\mathbf{r}/r^3$. Hence we obtain

$$
\operatorname{curl} \mathbf{A} = I_2 \oint_{C_2} d\mathbf{r}_2 \wedge \mathbf{r}/r^3 = \mathbf{H}
$$

as required. Also from eqn. (103)

$$
\begin{aligned}
\operatorname{div} \mathbf{A} &= I_2 \oint_{C_2} \operatorname{div}_r (d\mathbf{r}_2/r) \\
&= I_2 \oint_{C_2} [r^{-1} \operatorname{div}_r d\mathbf{r}_2 + d\mathbf{r}_2.\operatorname{grad}_r (r^{-1})] \\
&= I_2 \oint_{C_2} d\mathbf{r}_2.\operatorname{grad}_r r^{-1}
\end{aligned}
$$

since $\text{div}_r \, d\mathbf{r}_2 = 0$. Hence from Stokes' theorem

$$\text{div } \mathbf{A} = I_2 \int_{S_2} \text{curl}_{r_2} \, \text{grad}_r \, (r^{-1}) . d\mathbf{S}_2$$

$$= -I_2 \int_{S_2} \text{curl}_r \, \text{grad}_r \, (r^{-1}) . d\mathbf{S}_2 = 0 \quad (268a)$$

as required, since $\text{curl grad} \equiv 0$, where S_2 is any surface bounded by C_2.

9.10 CONTINUOUS CURRENT DISTRIBUTIONS

So far we have considered the magnetic fields due to currents in infinitely narrow wires. If, however, we have a continuous distribution of charge flow, specified by current density $\mathbf{J}$ (eqn. (262)), it follows from eqn. (264) that for circuit C_2

$$I_2 \, d\mathbf{r}_2 = J_2 \, dS_2 \, d\mathbf{r}_2 = \mathbf{J}_2 \, d\tau_2. \quad (269)$$

Hence we obtain from eqns. (265) and (268) that the magnetic field $\mathbf{H}$ and potential $\mathbf{A}$ for such a distribution is given by

$$\mathbf{H} = \int_\tau (\mathbf{J}_2 \wedge \mathbf{r}/r^3) \, d\tau_2 \quad (270)$$

$$\mathbf{A} = \int_\tau (\mathbf{J}_2/r) \, d\tau_2. \quad (271)$$

As before, these vectors satisfy the relations

$$\text{div } \mathbf{H} = 0, \quad \mathbf{H} = \text{curl } \mathbf{A}, \quad \text{div } \mathbf{A} = 0.$$

We can now proceed to prove the result that $\text{curl } \mathbf{H} = 4\pi\mathbf{J}$. We have

$$\text{curl } \mathbf{H} = \text{curl curl } \mathbf{A} = \text{grad div } \mathbf{A} - \nabla^2\mathbf{A} = -\nabla^2\mathbf{A} \quad (272)$$

from eqn. (117), since $\text{div } \mathbf{A} = 0$. Also if we write eqn. (271) in component form, we have

$$A_x = \int_\tau (J_{2x}/r) \, d\tau_2, \quad A_y = \int_\tau (J_{2y}/r) \, d\tau_2, \quad A_z = \int_\tau (J_{2z}/r) \, d\tau_2,$$

and it is seen that each of these equations is of the form

$$V(\mathbf{r}) = \int_\tau (\rho/\varepsilon_0 r) \, d\tau_2 \quad (273)$$

if $V = A$ and $(\rho/\varepsilon_0) = J_2$. Now eqn. (273) gives the *electrostatic* potential V at $\mathbf{r}$ due to a *charge* density ρ throughout the volume τ and we have shown in §9.3 that V as given by eqn. (273) satisfies the equation $\nabla^2 V = -4\pi\rho/\varepsilon_0$. Thus if we let $J_{2x}, J_{2y}, J_{2z} = \rho/\varepsilon_0$ in turn, it follows that A_x, A_y, A_z satisfy the equations

$$\nabla^2 A_x = -4\pi J_{2x}, \quad \nabla^2 A_y = -4\pi J_{2y}, \quad \nabla^2 A_z = -4\pi J_{2z}.$$

Hence

$$\nabla^2 \mathbf{A} = \mathbf{i}\nabla^2 A_x + \mathbf{j}\nabla^2 A_y + \mathbf{k}\nabla^2 A_z$$

$$= -4\pi(\mathbf{i}J_{2x} + \mathbf{j}J_{2y} + \mathbf{k}J_{2z}) = -4\pi\mathbf{J},$$

and substituting this in eqn. (272) gives

$$\text{curl } \mathbf{H} = -\nabla^2 \mathbf{A} = 4\pi\mathbf{J}. \tag{274}$$

Combining eqns. (266) and (269) it is seen that for the case of a continuous current distribution, the force $\delta\mathbf{F}$ acting on an element $\delta\tau$ with current density $\mathbf{J}$ due to an applied magnetic field $\mathbf{H}$ is given by

$$\delta\mathbf{F} = \mu_0 \mathbf{J} \wedge \mathbf{H}\, \delta\tau. \tag{275}$$

Consider now $\oint_C \mathbf{H}.d\mathbf{r}$ around any closed circuit C. We have from Stokes' theorem

$$\oint_C \mathbf{H}.d\mathbf{r} = \int_s \text{curl } \mathbf{H}.d\mathbf{S} = 4\pi \int_s \mathbf{J}.d\mathbf{S}$$

from eqn. (274). Making use of expression (264) for the total current I enclosed by the circuit C, this yields

$$\oint \mathbf{H}.d\mathbf{r} = 4\pi I.$$

9.11 ENERGY OF THE MAGNETIC FIELD

Consider a circuit C_1 in two positions A and B separated by $\delta\mathbf{r}$ as shown in Fig. 69. Then the force on an element $\delta\mathbf{r}_1$ of the circuit is $\delta\mathbf{F} = \mu_0 I_1\, \delta\mathbf{r}_1 \wedge \mathbf{H}$ (see eqn. (266)), and so the work dW done on the element $\delta\mathbf{r}_1$ in moving it from A to B is given by

$$dW = \delta\mathbf{r}.\delta\mathbf{F} = \mu_0 I_1 \, \delta\mathbf{r}.(\delta\mathbf{r}_1 \wedge \mathbf{H})$$
$$= \mu_0 I_1 \mathbf{H}.(\delta\mathbf{r} \wedge \delta\mathbf{r}_1) \qquad (276)$$

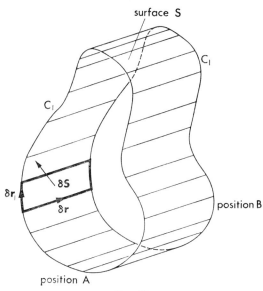

FIG. 69.

as the triple scalar product is invariant to cyclic permutation. Now, if δS is an element of the surface S mapped out by the circuit C_1 in its movement from A to B, it is clear that $\delta\mathbf{S} = \delta\mathbf{r} \wedge \delta\mathbf{r}_1$, and so eqn. (276) yields $dW = \mu_0 I_1 \mathbf{H}.\delta\mathbf{S}$, which on integrating for the whole circuit gives

$$\delta W = \mu_0 I_1 \int_s \mathbf{H}.d\mathbf{S}. \qquad (277)$$

We define now the magnetic flux Φ for the circuit C_1 in a given position by

$$\Phi = \mu_0 \int_{s_1} \mathbf{H}.d\mathbf{S}_1 \qquad (277a)$$

where S_1 is a surface bounded by C_1. It follows then from the divergence theorem that since div $\mathbf{H} = 0$, $\int \mathbf{H}.d\mathbf{S} = 0$, when this integral is taken over the closed surface bounded by S and the surfaces S_{1A}, S_{1B} enclosed by the circuit in its two positions A and B. Hence from the above definition (277) of Φ, we see that

$$\mu_0 \int_s \mathbf{H}.d\mathbf{S} = \Phi_B - \Phi_A\dagger = \delta\Phi \ .$$

for $\delta\mathbf{r}$ small. Substituting this into eqn. (277) gives

$$\delta W = I_1 \, \delta\Phi$$

showing that the energy of the circuit in the magnetic field is given by the function

$$W = I_1\Phi \tag{278}$$

assuming that I_1 is maintained constant throughout. Since $\mathbf{H} = \text{curl } \mathbf{A}$, it follows from Stokes' theorem that

$$\Phi = \mu_0 \int_{S_1} \text{curl } \mathbf{A}.d\mathbf{S}_1 = \mu_0 \oint_{C_1} \mathbf{A}.d\mathbf{r}_1. \tag{279}$$

Consider now the energy of a continuous current distribution $\mathbf{J}(\mathbf{r})$ placed in a magnetic field. Combining eqns. (278) and (279) and considering the continuous distributions as a set of current filaments, we have

$$W = \mu_0 \int_s \left(J \oint_{C_1} \mathbf{A}.d\mathbf{r}_1 \right) dS = \mu_0 \int \mathbf{A}.\mathbf{J} \, d\tau \tag{280}$$

since $I = \int J \, dS$, $\mathbf{J}$ is parallel to $d\mathbf{r}_1$ and $d\tau = dr_1 \, dS$. This is the energy when the current distribution is placed into an

---

† The reason for the minus sign in this term is that area elements δS_1 of S_{1A} and S_{1B} will have their normals in *opposite* directions when giving contributions to

$$\int_{s_{1A} + s_{1B} + s} \mathbf{H}.d\mathbf{S},$$

while their normals will be in the *same* direction when they are employed in eqn. (277a).

already existing magnetic field. If, however, we require the energy of the current distribution in the field that it sets up itself, then it may be shown by a similar argument to that used in §9.4 for the electrostatic case that the above result (280) must be multiplied by a factor of one half to yield

$$W = \tfrac{1}{2}\mu_0 \int \mathbf{A} . \mathbf{J} \, d\tau. \tag{281}$$

Using eqns. (274) and (203) we then have

$$W = \frac{\mu_0}{8\pi} \int \mathbf{A} . \operatorname{curl} \mathbf{H} \, d\tau = \frac{\mu_0}{8\pi} \left[\int \mathbf{H} . \operatorname{curl} \mathbf{A} \, d\tau - \int \operatorname{div} (\mathbf{A} \wedge \mathbf{H}) \, d\tau \right].$$

The second integral becomes $\int \mathbf{A} \wedge \mathbf{H} . d\mathbf{S}$ which vanishes over the boundary surface at infinity, and since $\mathbf{H} = \operatorname{curl} \mathbf{A}$, the the first term yields

$$W = (\mu_0/8\pi) \int H^2 \, d\tau, \tag{282}$$

which may be compared with the result (248).

9.12   ELECTROMAGNETIC INDUCTION

So far we have been concerned with static charges and steady currents. We shall now proceed to discuss the effects arising when the electrical quantities involved are time-dependent. The first new effect appearing under these conditions is the phenomenon of electromagnetic induction which states that if the magnetic flux Φ through a closed circuit is changing with time, then a potential difference V is set up around the circuit, which is given by

$$V = -\partial\Phi/\partial t. \tag{283}$$

This law may be considered as a basic experimental result, or alternatively it can be derived on the basis of energy conservation from eqn. (278) assuming that this equation remains unchanged if the flux alters with time.

Let us now apply eqn. (283) to a (imaginary) circuit C in a continuous current distribution. Then $V = \oint_C \mathbf{E}.\,d\mathbf{r}$, where $\mathbf{E}$ is the electric field set up in the circuit, and from eqn. (277) $\Phi = \int_s \mathbf{B}.\,d\mathbf{S}$, where S is a surface bounded by the circuit, and $\mathbf{B} = \mu_0\mathbf{H}$. Substituting these expressions into eqn. (283) then gives

$$\int_s \text{curl}\,\mathbf{E}.\,d\mathbf{S} = \oint_C \mathbf{E}.\,d\mathbf{r} = V = -\partial\Phi/\partial t = -\int(\partial\mathbf{B}/\partial t).\,d\mathbf{S}.$$

Since this is true for any arbitrary circuit C, it follows that

$$\text{curl}\,\mathbf{E} = -\partial\mathbf{B}/\partial t = -\mu_0\,\partial\mathbf{H}/\partial t. \tag{284}$$

Hence the effect of a changing magnetic field is to substitute eqn. (284) for the result $\text{curl}\,\mathbf{E} = 0$ derived in §9.1 for the case of electrostatics. Since in the presence of a changing magnetic field $\text{curl}\,\mathbf{E}$ is non-zero, it follows that $\mathbf{E}$ cannot now be derived from a scalar potential V via $\mathbf{E} = -\text{grad}\,V$.

9.13   THE DISPLACEMENT CURRENT

It was shown in §9.10 that for steady currents, the relation $\text{curl}\,\mathbf{H} = 4\pi\mathbf{J}$ holds. However, it is easy to show that this result cannot be true when the electrical quantities are time-dependent, for it implies that

$$\text{div}\,\mathbf{J} = (1/4\pi)\,\text{div curl}\,\mathbf{H} = 0, \tag{285}$$

while it has previously been shown in §9.7 that, in fact,

$$\text{div}\,\mathbf{J} = -(\partial\rho/\partial t). \tag{286}$$

Clearly eqns. (285) and (286) can only hold simultaneously if ρ is time-independent. How then should eqn. (274) be modified to deal with time-dependence? There are here no experimental results which show directly how to effect the required modification. Instead we pursue the following plausible line of thought, the ultimate justification of which is that results to be deduced later from the modified eqn. (289) agree with experiment. It is reasonable that the required modification of $\mathbf{J} = (1/4\pi)$ curl $\mathbf{H}$ should take the form

$$\mathbf{J} = (1/4\pi) \text{ curl } \mathbf{H} + \mathbf{M} \tag{287}$$

where $\mathbf{M}$ is a vector to be chosen such that the form (287) for $\mathbf{J}$ satisfies eqn. (286). Substituting therefore from eqn. (287) into eqn. (286) gives

$$\text{div } \mathbf{J} = \text{div } \mathbf{M} = -(\partial\rho/\partial t). \tag{288}$$

We would like $\mathbf{M}$ to depend only on the existing fields (electric or magnetic), and this can be achieved by substituting for ρ in eqn. (288) from eqn. (239) (for free space) to give

$$\text{div } \mathbf{M} = -(\partial\rho/\partial t) = -(\varepsilon_0/4\pi) \text{ div } (\partial\mathbf{E}/\partial t),$$

whence $\mathbf{M} = -(\varepsilon_0/4\pi)(\partial\mathbf{E}/\partial t)$. Hence eqn. (287) becomes

$$\text{curl } \mathbf{H} = 4\pi\mathbf{J} + \varepsilon_0(\partial\mathbf{E}/\partial t), \tag{289}$$

and this equation is the required generalisation of eqn. (274). The new term $\varepsilon_0(\partial\mathbf{E}/\partial t)$ appearing on the R.H.S. of eqn. (289) is called the displacement current.

9.14   MAXWELL'S EQUATIONS

Restricting our considerations to free space, we now see from eqns. (239), (284), (266), (289), that the electric and magnetic fields $\mathbf{E}$ and $\mathbf{H}$ satisfy the four Maxwell equations

$$\text{div } \mathbf{E} = 4\pi\rho/\varepsilon_0, \tag{290a}$$

$$\text{curl } \mathbf{E} = -\mu_0(\partial\mathbf{H}/\partial t), \tag{290b}$$

$$\text{div } \mathbf{H} = 0, \tag{290c}$$

$$\text{curl } \mathbf{H} = 4\pi\mathbf{J} + \varepsilon_0(\partial\mathbf{E}/\partial t). \tag{290d}$$

It also follows from eqns. (233a), (266), (269), (248), (282) that the force $\delta\mathbf{F}$ on an element $\delta\tau$ is given by

$$\delta\mathbf{F} = (\rho\mathbf{E} + \mu_0\mathbf{J} \wedge \mathbf{H})\,\delta\tau, \tag{291}$$

and that the energy δK of an element $\delta\tau$ is given by

$$\delta K = (1/8\pi)(\varepsilon_0 E^2 + \mu_0 H^2)\,\delta\tau. \tag{292}$$

We shall now show that the four Maxwell equations, together with suitable boundary conditions in space and time, yield unique solutions for $\mathbf{E}$ and $\mathbf{H}$. For space boundary conditions we shall suppose that both $\mathbf{E}$ and $\mathbf{H}$ are of the order $(1/r^2)$ for large r, and as time boundary condition we shall suppose that $\mathbf{E}$ and $\mathbf{H}$ are given for all $\mathbf{r}$ at $t = 0$. Then if there exist two solutions $\mathbf{E}_1$ and $\mathbf{E}_2$, $\mathbf{H}_1$ and $\mathbf{H}_2$ for $\mathbf{E}$, $\mathbf{H}$ respectively, let $\mathbf{F} = \mathbf{E}_1 - \mathbf{E}_2$ and $\mathbf{G} = \mathbf{H}_1 - \mathbf{H}_2$. Then $\mathbf{F}$ and $\mathbf{G}$ satisfy

(1) $\operatorname{div}\mathbf{F} = 0,$ (3) $\operatorname{div}\mathbf{G} = 0,$

(2) $\operatorname{curl}\mathbf{F} = -\mu_0(\partial\mathbf{G}/\partial t),$ (4) $\operatorname{curl}\mathbf{G} = \varepsilon_0(\partial\mathbf{F}/\partial t),$

together with the same space boundary conditions above, and the time boundary condition, $\mathbf{F} = \mathbf{G} = 0$ for $t = 0$. Then from eqns. (2) and (4) we have

$$\mathbf{G}.\operatorname{curl}\mathbf{F} = -\mu_0\mathbf{G}.(\partial\mathbf{G}/\partial t),$$
$$\mathbf{F}.\operatorname{curl}\mathbf{G} = \varepsilon_0\mathbf{F}.(\mathrm{d}\mathbf{F}/\mathrm{d}t)$$

which on subtracting yield

$$\operatorname{div}(\mathbf{G}\wedge\mathbf{F}) = \mathbf{F}.\operatorname{curl}\mathbf{G} - \mathbf{G}.\operatorname{curl}\mathbf{F}$$
$$= \varepsilon_0\mathbf{F}.(\partial\mathbf{F}/\partial t) + \mu_0\mathbf{G}.(\partial\mathbf{G}/\partial t)$$
$$= \tfrac{1}{2}\partial(\varepsilon_0 F^2 + \mu_0 G^2)/\partial t \tag{293}$$

making use of eqn. (183). Integrating both sides of eqn. (293) over all space then yields

$$(\mathrm{d}/\mathrm{d}t)\int_\tau (\varepsilon_0 F^2 + \mu_0 G^2)\,\mathrm{d}\tau = 2\int \operatorname{div}(\mathbf{G}\wedge\mathbf{F})\,\mathrm{d}\tau$$
$$= 2\int (\mathbf{G}\wedge\mathbf{F}).\mathrm{d}\mathbf{S} = 0 \tag{294}$$

since $\mathbf{G}$ and $\mathbf{F}$ vanish sufficiently rapidly on the boundary surface at infinity. Now since $\mathbf{F} = \mathbf{G} = 0$ for all $\mathbf{r}$ at $t = 0$ (the time boundary condition), it follows that $I \equiv \int_\tau (\varepsilon_0 F^2 + \mu_0 G^2) \, d\tau = 0$ at $t = 0$, and since we have shown in eqn. (294) that $dI/dt = 0$, it follows that $I = 0$ for all t. Hence since F^2 and G^2 are necessarily positive or zero, we see that $\mathbf{F} = \mathbf{G} = 0$ for all $\mathbf{r}$ and t. Thus $\mathbf{E}_1 = \mathbf{E}_2$ and $\mathbf{H}_1 = \mathbf{H}_2$, showing that the Maxwell equations have a unique solution.

9.15   THE ELECTROMAGNETIC POTENTIALS

In the previous section we considered the four Maxwell equations which determine the electric and magnetic fields in free space. For some purposes, however, it is more convenient to work in terms of the scalar and vector potentials V and $\mathbf{A}$. We have previously discussed these for the case of static charges and steady currents, and we shall now consider them for time-dependent fields, starting from Maxwell's equations. It follows from eqn. (290c) that a vector $\mathbf{A}$ exists such that

$$\mathbf{H} = \operatorname{curl} \mathbf{A} \tag{295}$$

and substituting this into eqn. (290b) gives

$$\operatorname{curl} [\mathbf{E} + \mu_0(\partial \mathbf{A}/\partial t)] = 0,$$

whence it follows that a scalar V exists such that

$$\mathbf{E} + \mu_0(\partial \mathbf{A}/\partial t) = -\operatorname{grad} V. \tag{296}$$

Substituting then from eqns. (296) and (290d) into eqns. (290a) and (295) yields

$$-\nabla^2 V - \mu_0 \operatorname{div}(\partial \mathbf{A}/\partial t) = 4\pi\rho/\varepsilon_0 \tag{297}$$

and

$$\operatorname{curl} \operatorname{curl} \mathbf{A} = \operatorname{curl} \mathbf{H} = 4\pi \mathbf{J} - \varepsilon_0\mu_0 \, \partial^2 \mathbf{A}/\partial t^2 - \varepsilon_0 \, \partial(\operatorname{grad} V)/\partial t;$$

i.e.

$$-\nabla^2 \mathbf{A} + \varepsilon_0\mu_0 \, \partial^2 \mathbf{A}/\partial t^2 + \operatorname{grad} [\operatorname{div} \mathbf{A} + \varepsilon_0(\partial V/\partial t)] = 4\pi \mathbf{J} \tag{298}$$

from eqn. (117). Now we have already specified curl $\mathbf{A}(= \mathbf{H})$, but in order to define $\mathbf{A}$ uniquely, div $\mathbf{A}$ must be given independently as discussed in §5.6. We shall let

$$\text{div } \mathbf{A} = -\varepsilon_0(\partial V/\partial t), \tag{299}$$

when eqns. (297) and (298) become

$$-\nabla^2 V + \varepsilon_0\mu_0(\partial^2 V/\partial t^2) = 4\pi\rho/\varepsilon_0 \tag{300}$$

and

$$-\nabla^2 \mathbf{A} + \varepsilon_0\mu_0(\partial^2 \mathbf{A}/\partial t^2) = 4\pi\mathbf{J} \tag{301}$$

respectively. The three eqns. (299), (300) and (301) determine V and $\mathbf{A}$, for from which $\mathbf{H}$ and $\mathbf{E}$ may be obtained via eqns. (295) and (296). The eqns. (299), (300), (301) for V and $\mathbf{A}$ differ from those obtained earlier (268a, 240, 274) for time-independent fields only by virtue of the terms involving $\partial/\partial t$ and $\partial^2/\partial t^2$.

9.16 ELECTROMAGNETIC WAVES

Let us consider now Maxwell's equations in empty space for which $\rho = \mathbf{J} = 0$. Then

$$\text{div } \mathbf{E} = 0 \tag{302a}$$

$$\text{curl } \mathbf{E} = -\mu_0(\partial \mathbf{H}/\partial t) \tag{302b}$$

$$\text{div } \mathbf{H} = 0 \tag{302c}$$

$$\text{curl } \mathbf{H} = \varepsilon_0(\partial \mathbf{E}/\partial t). \tag{302d}$$

From eqns. (302d), (302b), (117) and (302a), we have

$$\varepsilon_0(\partial^2 \mathbf{E}/\partial t^2) = (\partial/\partial t)\,\text{curl }\mathbf{H} = \text{curl}\,(\partial \mathbf{H}/\partial t)$$

$$= -(1/\mu_0)\,\text{curl curl }\mathbf{E}$$

$$= -(1/\mu_0)\,(\text{grad div }\mathbf{E} - \nabla^2\mathbf{E}) = (1/\mu_0)\nabla^2\mathbf{E}.$$

Thus

$$(\partial^2 \mathbf{E}/\partial t^2) = (1/\varepsilon_0\mu_0)\nabla^2\mathbf{E}. \tag{303a}$$

Similarly, on working from eqn. (302b) we obtain

$$\partial^2 \mathbf{H}/\partial t^2 = (1/\varepsilon_0 \mu_0)\nabla^2 \mathbf{H}, \tag{303b}$$

showing that $\mathbf{H}$ and $\mathbf{E}$ satisfy the same equation. If $\mathbf{F} = \mathbf{E}$ or $\mathbf{H}$ and $c^2 = (1/\varepsilon_0 \mu_0)$, then $\mathbf{E}$ and $\mathbf{H}$ each satisfy the equation

$$\partial^2 \mathbf{F}/\partial t^2 = c^2 \nabla^2 \mathbf{F}. \tag{304}$$

The detailed solution of this equation depends on the boundary conditions, but in all cases the variation of $\mathbf{F}$ in space and time can be shown to correspond to a wave motion which is propagated with velocity c. We shall now show this for the special case where $\mathbf{F}$ does not vary with x or y, and therefore eqn. (304) becomes

$$\partial^2 \mathbf{F}/\partial t^2 = c^2 \partial^2 \mathbf{F}/\partial z^2. \tag{305}$$

Now, if the time and space variation of $\mathbf{F}$ is to correspond to a wave motion propagated in the z direction with velocity c,

$$\mathbf{F} = \mathbf{F_0} f(z \pm ct)$$

where $\mathbf{F_0}$ is a constant vector and f is any function. Then

$$\partial^2 \mathbf{F}/\partial z^2 = \mathbf{F_0}(\mathrm{d}^2 f/\mathrm{d}u^2)$$

where $u = z \pm ct$, and

$$\partial^2 \mathbf{F}/\partial t^2 = c^2 \mathbf{F_0}(\mathrm{d}^2 f/\mathrm{d}u^2).$$

Hence it follows that eqn. (305) is satisfied. The electromagnetic waves which occur as the solutions of eqn. (304) give rise to light waves and radio waves, the distinction between the two types of wave being due to their different wavelengths.

Propagation of Energy

Finally, we consider the propagation of energy by the electromagnetic waves whose electric and magnetic fields satisfy the eqns. (302a) to (302d). For an arbitrary volume τ, surrounded by a surface S, the total electromagnetic energy K enclosed is given by eqn. (292) as

$$K = (1/8\pi) \int_\tau (\varepsilon_0 \mathbf{E}^2 + \mu_0 \mathbf{H}^2)\, d\tau.$$

Hence

$$(dK/dt) = (1/4\pi) \int_\tau [\varepsilon_0 \mathbf{E}.(\partial \mathbf{E}/\partial t) + \mu_0 \mathbf{H}.(\partial \mathbf{H}/\partial t)]\, d\tau$$

$$= (1/4\pi) \int_\tau (\mathbf{E}.\operatorname{curl} \mathbf{H} - \mathbf{H}.\operatorname{curl} \mathbf{E})\, d\tau$$

making use of eqns. (302b) and (302d). Thus from eqn. (103)

$$dK/dt = -(1/4\pi) \int_\tau \operatorname{div}(\mathbf{E} \wedge \mathbf{H})\, d\tau$$

$$= -(1/4\pi) \int_s (\mathbf{E} \wedge \mathbf{H}).d\mathbf{S}. \tag{306}$$

Hence $(1/4\pi) \int (\mathbf{F} \wedge \mathbf{H}).d\mathbf{S}$ is the net rate at which energy is leaving the volume τ, and so the vector

$$\mathbf{P} = (1/4\pi)(\mathbf{E} \wedge \mathbf{H}) \tag{307}$$

may be interpreted as giving the rate of flow of energy per unit area at any point. $\mathbf{P}$ is generally known as Poynting's vector.